DESIGN AND CONSTRUCTION OF BERM BREAKWATERS

ADVANCED SERIES ON OCEAN ENGINEERING

Series Editor-in-Chief
Philip L-F Liu (*Cornell University*)

Published

Vol. 29 Coastal and Estuarine Processes
by Peter Nielsen (The Univ. of Queensland, Australia)

Vol. 30 Introduction to Coastal Engineering and Management (2nd Edition)
by J. William Kamphuis (Queen's Univ., Canada)

Vol. 31 Japan's Beach Erosion: Reality and Future Measures
by Takaaki Uda (Public Works Research Center, Japan)

Vol. 32 Tsunami: To Survive from Tsunami
by Susumu Murata (Coastal Development Inst. of Technology, Japan),
Fumihiko Imamura (Tohoku Univ., Japan),
Kazumasa Katoh (Musashi Inst. of Technology, Japan),
Yoshiaki Kawata (Kyoto Univ., Japan),
Shigeo Takahashi (Port and Airport Research Inst., Japan) *and*
Tomotsuka Takayama (Kyoto Univ., Japan)

Vol. 33 Random Seas and Design of Maritime Structures (3rd Edition)
by Yoshimi Goda (Yokohama National University, Japan)

Vol. 34 Coastal Dynamics
by Willem T. Bakker (Delft Hydraulics, Netherlands)

Vol. 35 Dynamics of Water Waves: Selected Papers of Michael Longuet-Higgins Volumes 1–3
edited by S. G. Sajjadi (Embry-Riddle Aeronautical University, USA)

Vol. 36 Ocean Surface Waves: Their Physics and Prediction (2nd Edition)
by Stanislaw R. Massel (Institute of Oceanology of the Polish Academy of Sciences, Sopot, Poland)

Vol. 37 Computational Wave Dynamics
by Hitoshi Gotoh (Kyoto University, Japan),
Akio Okayasu (Tokyo University of Marine Science and Technology, Japan)
and Yasunori Watanabe (Hokkaido University, Japan)

Vol. 38 An Introduction to Hydraulics of Fine Sediment Transport
by Ashish J. Mehta (University of Florida, USA)

Vol. 39 Liquefaction Around Marine Structures
(With CD-ROM)
by Mutiu Sumer (Technical University of Denmark, Denmark)

Vol. 40 Design and Construction of Berm Breakwaters
by Van der Meer, Jentsje (Van der Meer Consulting BV, The Netherlands & UNESCO-IHE, The Netherlands) *and*
Sigurdur Sigurdarson (IceBreak Consulting Engineers ehf, Iceland)

*For the complete list of titles in this series, please write to the Publisher.

Advanced Series on Ocean Engineering — Volume 40

DESIGN AND CONSTRUCTION OF BERM BREAKWATERS

Jentsje van der Meer
Van der Meer Consulting BV & UNESCO-IHE, the Netherlands

Sigurdur Sigurdarson
IceBreak Consulting Engineers ehf, Iceland

World Scientific

NEW JERSEY • LONDON • SINGAPORE • BEIJING • SHANGHAI • HONG KONG • TAIPEI • CHENNAI • TOKYO

Published by

World Scientific Publishing Co. Pte. Ltd.

5 Toh Tuck Link, Singapore 596224

USA office: 27 Warren Street, Suite 401-402, Hackensack, NJ 07601

UK office: 57 Shelton Street, Covent Garden, London WC2H 9HE

Library of Congress Cataloging-in-Publication Data
Names: Van der Meer, Jentsje, author. | Sigurdarson, Sigurdur, author.
Title: Design and construction of berm breakwaters / Van der Meer, Jentsje
(Van der Meer Consulting BV, Netherlands, UNESCO-IHE, Netherlands),
Sigurdur Sigurdarson (IceBreak Consulting Engineers ehf, Iceland).
Description: [Hackensack] New Jersey : World Scientific, [2016] |
Includes bibliographical references.
Identifiers: LCCN 2016015372 | ISBN 9789814749602 (hardback : alk. paper)
Subjects: LCSH: Rubble mound breakwaters--Design and construction.
Classification: LCC TC333 .M36 2016 | DDC 627/.24--dc23
LC record available at https://lccn.loc.gov/2016015372

British Library Cataloguing-in-Publication Data
A catalogue record for this book is available from the British Library.

Photo on the cover: The berm breakwater at Arnarstapi, Iceland, sheltering a small fishing harbour. Photo Mats Wibe Lund.

Desk Editor: Amanda Yun

Typeset by Stallion Press
Email: enquiries@stallionpress.com

Printed in Singapore

Preface

The scientific experience with berm breakwaters of Dr Jentsje van der Meer, and the practical experience of Sigurdur Sigurdarson in the design and construction of over thirty berm breakwaters all over the world, come together in this book. While written for the (practical) designer, it also includes a scientific background as validation, which may be of interest to hydraulic modellers and researchers.

Whilst some mass-armoured breakwaters have been built before, the modern design of so-called “berm breakwaters” started with designs in 1983 by Baird & Associates in Ottawa. The original design consisted of mass-armoured berms that were expected to reshape to statically stable S-shaped slopes. The design was adopted in Iceland and eventually led to a development with more stable structures by utilizing available rock sizes, large rock and more gradings. This more stable and only partly reshaping structure is called the Icelandic-type berm breakwater.

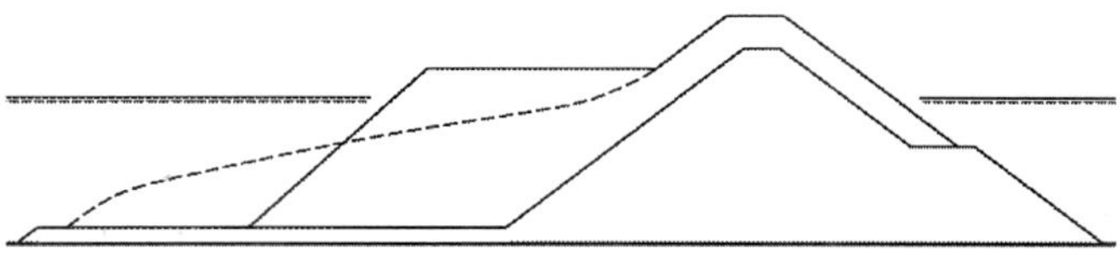

Original design of modern berm breakwaters.

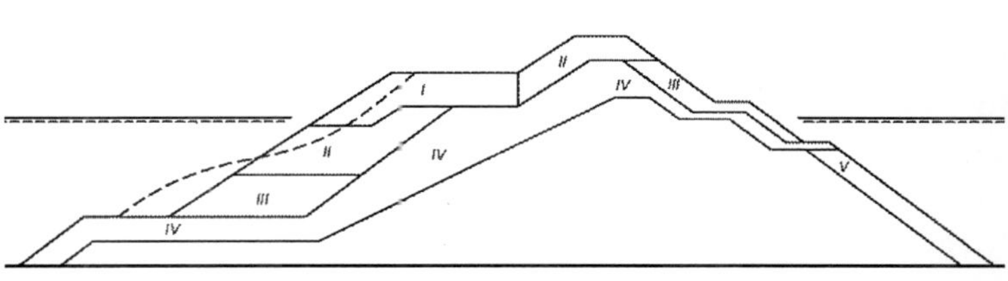

Icelandic-type berm breakwater; more rock classes, less reshaping.

A new classification of berm breakwaters has been given in this book. There are three classes to describe the behaviour of the berm: hardly reshaping, (HR), partly reshaping, (PR), and fully reshaping, (FR). The first two can be described as statically stable (description mainly by damage and some recession). The fully reshaping berm breakwater is potentially unstable directly after construction, but the reshaped profile is statically stable. This means that under design conditions, the stability number $H_s/\Delta D_{n50}$ should not exceed 3. The Icelandic-type berm breakwater falls into the first two classes, the mass-armoured one in the last two classes. The classification gives stability numbers, damage and recession for each type of berm breakwater.

Stability formulae for conventional rock layers under wave attack can be applied for estimating damage for hardly reshaping and partly reshaping berm breakwaters. A new method has been described for prediction of the recession of each type of berm breakwater. This prediction method uses the stability number $H_s/\Delta D_{n50}$ as basis and the influence of wave period has been proven to be hardly or non-existing. Also, geometrical aspects as lower slope, berm level and toe depth on influence of recession have been described. Finally, the functional behaviour of berm breakwaters, like wave overtopping, reflection and transmission have been treated with new design formulae as a result.

"Rock" is the essential term in the design of berm breakwaters. Often, dedicated quarries can be found and opened to produce the required rock. This is different from designs with rock demand from existing quarries, where delivery of very large rock classes may be problematic. It has been proven possible to win really large rock in dedicated quarries and all this experience has been described in the book. This does however require changes to quarry investigation methods and in blasting. Quarry and project management as well as blasting and sorting techniques are essential in getting all required rock for an acceptable price.

Very often construction can be limited to the use of split barges for under water construction and large excavators for the placement of rock, even up to 30 t and more. Large cranes are not always required. The use of excavators enables specific placement of the large Class I rock above water, which increases stability against wave action.

The book is neither a research paper nor does it summarise all research that has been performed in the past. It uses the main results of earlier research to come to a description of the behaviour of berm breakwaters, and to arrive at practical ways of design and construction. Guidelines on the geometrical design of the cross-section of a berm breakwater have been described in Chapter 5. Such guidelines have been lacking until now, so this book may indeed replace sections 5.2.2.6 and 6.1.6 in the Rock Manual (2007).

Most of the work comes together in Chapter 8 where practical design guidance is given on how to design a berm breakwater for various design wave conditions and available rock classes. The final Chapter, Chapter 9, describes constructed examples of berm breakwaters, and the experience since their original construction.

Jentsje van der Meer and Sigurdur Sigurdarson at the new breakwater at Helguvik, Iceland.

Acknowledgment

Writing a book might be a dream of many scientists and researchers. But a decision to write a book when running one's own business means donating a lot of personal free time over a number of years. The idea for two authors to write a book together made the decision easier: each author had to carry only half the burden. Moreover, there is a lot of fun in traveling, discussing and sometimes writing together, in the Netherlands, Iceland, or even Curaçao. These trips have not only been valuable for the authors, but also for our wives. Writing this book became a pleasure in this way and made it acceptable for all four of us.

Some of the new aspects developed during the writing of this book have already been presented at conferences. We thank all our colleagues and co-authors for the discussion and feedback on those presentations. We thank our clients who involved us in the design and construction of berm breakwaters all over the world, which gave the basis for this book.

Designing breakwaters requires team work. We have been part of many teams over the years and we would like to thank our colleagues, co-workers and contractors for the cooperation. A special thanks is due to Thomas Lykke Andersen, who was open to share all his data with us and was willing to discuss details. Finally, we are grateful to William Allsop, who gave valuable comments on the contents of the book and prevented us from making too obvious mistakes in English grammar.

Contents

Preface ... v
Acknowledgment ... ix
Notation ... xv

1. History of Modern Berm Breakwaters ... 1
   1.1 Time before modern berm breakwaters ... 1
   1.2 Developments in Australia, mass-armoured breakwater ... 2
   1.3 Developments in Canada, modern berm breakwaters ... 3
   1.4 Contact between Canada and the Netherlands ... 5
   1.5 Developments in Iceland ... 6
   1.6 Berm breakwaters in international cooperation ... 12
   1.7 Outline of the book ... 13

2. Classification and Types of Berm Breakwaters ... 15
   2.1 Design wave climate, definitions and parameters ... 15
   2.2 Classification of berm breakwaters by PIANC ... 21
   2.3 New classification of berm breakwaters ... 22

3. Predicting Stability and Reshaping ... 29
   3.1 Practical aspects ... 29
   3.2 Existing prediction methods on static stability ... 30
   3.3 The modified Van der Meer formulae in the Rock Manual ... 35
   3.4 Existing prediction methods on reshaping ... 36
   3.5 Damage profile for a statically stable straight slope ... 43
   3.6 Validation of damage prediction ... 46
   3.7 New method for recession of berm breakwaters ... 52
      3.7.1 Available data sets ... 52
      3.7.2 Development of new recession formula ... 60

3.7.3 Influences on recession of berm breakwaters 66
3.7.4 Conclusions on recession of berm breakwaters 76

4. Functional Behaviour: Wave Overtopping, Reflection and Transmission 79
4.1 Wave overtopping 79
4.1.1 Context of wave overtopping 79
4.1.2 Available data sets 89
4.1.3 Development of influence factor γ_{BB} 92
4.2 Wave reflection 102
4.3 Wave transmission 107

5. Geometrical Design of the Cross-section 111
5.1 About design guidance in this chapter 111
5.2 Parameters in geometrical design of the cross-section 112
5.2.1 General description of the cross-section 112
5.2.2 Berm width, B, as function of recession and resiliency 115
5.2.3 Crest level, R_c 117
5.2.4 Horizontal armour width, A_h 120
5.2.5 Rock classes and proposal for new mass-armoured berm breakwater 121
5.2.6 Berm level, d_b 126
5.2.7 Apron 128
5.2.8 Transition from Class I to Class II rock 129
5.2.9 Possible toe berm 130
5.3 Application of geometrical rules for a design 131
5.4 Roundhead 135
5.5 Soft soil 142
5.6 Maintenance aspects 145

6. Armourstone and Quarrying 147
6.1 Introduction 147
6.2 Armourstone gradings 151
6.2.1 Introduction 151
6.2.2 Grading curves in research 151
6.2.3 EN 13383 system for standard gradings 153
6.2.4 Non-standard gradings from dedicated quarries 156
6.3 Quarry yield prediction 158
6.4 Blasting for armourstone 165
6.5 Quarry planning 170

7. Construction ... 173

7.1 Introduction ... 173

7.2 Armourstone quarry in the contract ... 173

7.3 Equipment ... 174

7.4 Placement and tolerances ... 180

7.4.1 Armourstone placement ... 180

7.4.2 Tolerances ... 185

7.4.3 Definition of rock surface and survey method for constructed profile ... 186

7.5 Weight assessment of rock gradings *in-situ* ... 190

7.5.1 Weight assessment by estimating average rock dimensions ... 190

7.5.2 Weight assessment of rock *in-situ* from block shape ... 193

7.5.3 Comparison of the two methods ... 196

8. Geometrical Design into Practice, Examples ... 199

8.1 Design methods ... 199

8.1.1 Geometrical design method for berm breakwaters ... 199

8.1.2 Design wave climate and other conditions for examples ... 208

8.2 Examples for a design wave height of 5 m ... 210

8.2.1 HR IC dedicated quarry, Class I 10-20 t ... 210

8.2.2 PR IC standard gradings, Class I 6-10 t ... 213

8.2.3 FR MA standard gradings, Class I 3-6 t ... 215

8.2.4 Conventional rock armour design ... 217

8.2.5 Overall conclusions and comparison ... 220

8.3 Examples for a design wave height of 3 m ... 221

8.3.1 HR IC dedicated quarry, Class I 1-4 t ... 221

8.3.2 PR MA dedicated quarry, Class I 0.5-2 t ... 223

8.3.3 Conventional rock armour design ... 225

8.3.4 Overall conclusions and comparison ... 227

8.4 Examples for design a wave height of 7 m ... 228

8.4.1 HR IC dedicated quarry, Class I 20-35 t ... 228

8.4.2 PR IC dedicated quarry, Class I 10-20 t ... 230

8.4.3 FR MA standard grading, Class I 6-10 t ... 232

8.4.4 Overall conclusions and comparison ... 234

9. Constructed Examples ... 237

9.1 Introduction ... 237

9.2 Hardly reshaping berm breakwater ... 237

9.2.1 The Dalvik berm breakwater, Iceland ... 237

9.2.2 The Helguvik berm breakwater, new design, Iceland 243
9.3 Partly reshaping berm breakwater - Icelandic-type 244
9.3.1 The Sirevåg berm breakwater, Norway 244
9.3.2 The Hammerfest berm breakwater, Norway 250
9.3.3 The Husavik berm breakwater, Iceland 253
9.4 Partly reshaping berm breakwater - mass-armoured 258
9.4.1 The Helguvik berm breakwater, original design, Iceland 258
9.5 Fully reshaping berm breakwater 261
9.5.1 The St George breakwaters, Alaska 261
9.5.2 The Bakkafjordur breakwater, Iceland 266
9.5.3 The Mortavika berm breakwater, Norway 276
9.5.4 The Mackay Southern breakwater, Queensland, Australia 282

Appendix A
Relationships to Compose a Damage Profile for a Straight Rock Slope 287

Appendix B
Detailed Analysis of Berm Recession 289

Appendix C
Detailed Analysis of Wave Overtopping 299

Appendix D
Calculations of Examples for Geometrical Design in Chapter 8 311

Bibliography 321

About the Authors 327

Notation

A	Coefficient in Eq. 5.7	(-)
A_e	Erosion area on rock profile	(m²)
A_h	Horizontal armour length, part of geometrical design of berm breakwater	(m)
B	Berm width of berm breakwater	(m)
B_{min}	Minimum berm width from geometrical criteria	(m)
BL_c	Blockiness, the volume of a block divided by the volume of the enclosing XYZ orthogonal	(-)
b	Width of a stone, middle dimension	(m)
b_{av}	Average width of a stone	(m)
D_{n15}	Nominal diameter, or equivalent cube size, $D_{n15} = (M_{15}/\rho_r)^{1/3}$	(m)
D_{n50}	Nominal diameter, or equivalent cube size, $D_{n50} = (M_{50}/\rho_r)^{1/3}$	(m)
D_{n85}	Nominal diameter, or equivalent cube size, $D_{n85} = (M_{85}/\rho_r)^{1/3}$	(m)
D_{nmax}	Nominal diameter, or equivalent cube size, of maximum stone size in a grading	(m)
D_{nmin}	Nominal diameter, or equivalent cube size, of minimum stone size in a grading	(m)
D_{50}	Sieve diameter, diameter of stone that exceeds the 50%-value of sieve curve	(m)
D_{85}	85% value of sieve curve	(m)
D_{15}	15% value of sieve curve	(m)
d_b	Level of berm of berm breakwater above DWL	(m)
G_c	Width of rock armoured crest, [Lykke Andersen, 2006]	(m)

g	Acceleration due to gravity (9.81)	(m/s^2)
H	Wave height, from trough to crest	(m)
$H_{1/3}$	Significant wave height based on time domain analysis, average of highest 1/3-rd of all wave heights	(m)
$H_{2\%}$	Wave height exceeded by 2% of waves	(m)
H_o	Stability number, $H_o = H_s/\Delta D_{n50}$	(-)
H_oT_o	Dynamic stability number, $H_oT_o = H_s/(\Delta D_{n50})\cdot T\cdot(g/D_{n50})^{0.5}$	(-)
H_oT_{om}	Dynamic stability number, $H_oT_{om} = H_s/(\Delta D_{n50})\cdot T_m\cdot(g/D_{n50})^{0.5}$	(-)
H_oT_{op}	Dynamic stability number, $H_oT_{op} = H_s/(\Delta D_{n50})\cdot T_p\cdot(g/D_{n50})^{0.5}$	(-)
H_0	Stability number, [Lykke Andersen, 2006], $H_0 = H_{m0}/\Delta D_{n50}$	(-)
H_0T_0	Dynamic stability number [Lykke Andersen, 2006] $H_0T_0 = H_{m0}/(\Delta D_{n50})\cdot T_{m0,1}\cdot(g/D_{n50})^{0.5}$	(-)
H_{m0}	Significant wave height calculated from the spectrum, $H_{m0} = 4\sqrt{m_0}$	(m)
$H_{m0,i}$	Incident significant wave height	(m)
$H_{m0,r}$	Reflected significant wave height	(m)
$H_{m0,t}$	Transmitted significant wave height	(m)
H_{max}	Maximum wave height in a record	(m)
H_s	Significant wave height, $H_s = H_{1/3}$	(m)
H_{sD}	Design wave height, taking the 100-years return period, or the second highest step in a research test series	(m)
$H_{s,i}$	Wave height of the i-th storm	(m)
$H_{2\%,i}$	2%-wave height of the i-th storm	(m)
$H_s/\Delta D_{n50}$	Stability number	(-)
$H_{sD}/\Delta D_{n50}$	Design stability number using H_{sD}	(-)
h	Water depth; water depth at structure toe	(m)
h	Height of a stone, smallest dimension	(m)
$h_{I\text{-}II}$	Transition on seaward slope from Class I to Class II measured from SWL	(m)
h_{av}	Average height of a stone	(m)
h_b	Point on a straight rock slope, vertically measured; part of the damage profile, see Figure 3.5	(m)

h_b	Level of berm of berm breakwater below SWL, [Lykke Andersen, 2006]	(m)
h_d	Point on a straight rock slope, vertically measured; part of the damage profile, see Figure 3.5	(m)
h_m	Point on a straight rock slope, vertically measured; part of the damage profile, see Figure 3.5	(m)
h_r	Point on a straight rock slope, vertically measured; part of the damage profile, see Figure 3.5	(m)
h_s	Step height in the reshaped profile, see Figure 3.4	(m)
h_t	Water depth above the toe berm	(m)
I_{s50}	Point load index	(Pa)
K_r	Reflection coefficient	(-)
K_t	Transmission coefficient	(-)
k	Wave number, $k = 2\pi/L_p$	(m^{-1})
L	Wave length, in the direction of propagation	(m)
L_o	Offshore or deep-water wave length, $L_o = gT^2/2\pi$	(m)
L_{om}	Offshore or deep-water wave length using the mean period, T_m	(m)
L_{op}	Offshore or deep-water wave length using the peak period, T_p	(m)
L_p	peak wave length calculated from the linear dispersion relation, [Lykke Andersen, 2006]	(m)
l_{av}	Average length of a stone	(m)
l	Length of a stone	(m)
M	Mass of stone	(kg)
M_0	Mass of particle for which 0% of the granular material is lighter	(kg)
M_5	Mass of particle for which 5% of the granular material is lighter	(kg)
M_{15}	Mass of particle for which 15% of the granular material is lighter	(kg)
M_{50}	Mass of particle for which 50% of the granular material is lighter	(kg)
M_{85}	Mass of particle for which 85% of the granular material is lighter	(kg)

M_{100}	Mass of particle for which 100% of the granular material is lighter	(kg)
M_{50min}	Minimum M_{50} that is possible in a gradinig	(kg)
M_{em}	Effective mean mass (of a standard grading), i.e. the arithmetic average of all pieces excluding those that fall below ELL for the grading	(kg)
M_{max}	Stone with maximum mass in a grading	(kg)
M_y	Mass of stone exceeded by y% on mass distribution curve	(kg)
M_{min}	Stone with minimum mass in a grading	(kg)
m_0	Zeroth moment of wave spectrum	(m^2s)
m_n	n-th moment of spectrum	($m^{2+n}s$)
N_{od}	Damage parameter. Number of stones displaced in a strip one D_{n50} wide	(-)
N_w	Number of waves over the duration of a storm record, or test, N_w = duration $/T_m$	(-)
$N_{w,i}$	Number of waves of the i-th storm	(-)
$N_{w,i\ i+1}$	Increased number of waves of the i+1-th storm, taking into account the damage caused by the i-th storm (to be used in cumulative damage calculation)	(-)
P	Notional permeability factor, defined by Van der Meer	(-)
$P_{\%}$	Part of berm that may reshape under design conditions; measure of resiliency	(%)
q	Time-averaged overtopping discharge per metre run of crest	(m^3/s per m)
R_c	Crest freeboard, level of crest relative to still water level	(m)
Rec	Recession of the berm of a berm breakwater	(m)
S_d	Non-dimensional damage parameter, $S_d = A_e / D_{n50}^2$ calculated from mean profiles or separately for each profile line, then averaged	(-)
S_{di}	S_d for i-th storm in a sequence of storms	(-)
s	Wave steepness, s = H/L	(-)
s_o	Fictitious wave steepness, defined as $H_s/L_o = 2\pi H_s/(gT_m^2)$	(-)
s_{om}	Fictitious wave steepness for mean period wave, $s_{om} = 2\pi H_s/(gT_m^2)$	(-)

s_{op}	Fictitious wave steepness for peak period wave, $s_{op} = 2\pi H_s /(gT_p^2)$	(-)
$s_{om\text{-}1,0}$	Fictitious wave steepness for mean energy period, $s_{m\text{-}1,0} = 2\pi H_{m0}/(gT_{m\text{-}1,0}^2)$	(-)
$s_{0,1}$	Fictitious wave steepness for mean period from spectrum [Lykke Andersen, 2006] $= 2\pi H_{m0}/(gT_{0,1}^2)$	(s)
T	Wave period	(s)
T_o	Wave period parameter for dynamic stability number H_oT_o, $T_o = T\cdot(g/D_{n50})^{0.5}$	(-)
T_{om}	Wave period parameter for dynamic stability number H_oT_o, $T_o = T_m\cdot(g/D_{n50})^{0.5}$	(-)
T_0	Wave period parameter for dynamic stability number H_0T_0, $T_0 = T_{0,1}\cdot(g/D_{n50})^{0.5}$	(-)
T_m	Mean wave period from time domain analysis	(s)
$T_{0,1}$	Mean wave period from spectrum [Lykke Andersen, 2006] $T_{0,1} = m_0/m_1$	(s)
$T_{m\text{-}1,0}$	Mean energy wave period or spectral wave period, $T_{m\text{-}1,0} = m_{-1}/m_0$	(s)
$T_{m,i}$	Mean wave period of the i-th storm	(s)
T_p	Spectral peak period, inverse of peak frequency	(s)
V_{rock}	Volume of a stone	(m^3)
X, Y, Z	Block dimensions of enclosing rectanguloid box with minimum volume, as used in blockiness calculation	(m)
α	Structure slope angle	(°)
α_d	Structure slope angle of the lower slope	(°)
α_u	Structure slope angle of the upper slope	(°)
β	Angle of wave attack with respect to the structure	(°)
γ_{BB}	Reduction factor for a berm of a berm breakwater (wave overtopping)	(-)
γ_b	Reduction factor for a berm in a sloping structure (wave overtopping)	(-)
γ_f	Reduction factor for slope roughness (wave overtopping)	(-)
γ_v	Reduction factor for a vertical wall on top of a sloping structure (wave overtopping)	(-)
γ_β	Reduction factor for oblique waves (wave overtopping)	(-)

Δ	Relative buoyant density of material, ie for rock $\Delta = \rho_r / \rho_w - 1$	(-)
Δw	Safe working level above MHWS	(m)
ξ	Surf similarity parameter, $\xi = \tan\alpha / \sqrt{s}$	(-)
ξ_m	Surf similarity parameter for mean wave period T_m	(-)
ξ_{cr}	Critical surf similarity parameter to distinguish between breaking and not breaking waves	(-)
$\xi_{m-1,0}$	Surf similarity parameter for spectral wave period $T_{m-1,0}$ and spectral significant wave height H_{m0}	(-)
ξ_p	Surf similarity parameter for peak wave period T_p	(-)
ρ_r	Mass density of rock	(kg/m³)
ρ_w	Mass density of water	(kg/m³)
$\sigma(x)$	Standard deviation	(Unit of x)

Chapter 1

History of Modern Berm Breakwaters

1.1 Time before modern berm breakwaters

Several nineteenth century breakwaters are usually reported as the origin of berm breakwaters, [Bruun and Johannesson, 1976], [Baird and Hall, 1983]. Many of these were exposed to waves too high in relation to the size of rock available for construction. Steep slopes resulted in severe damage, often repaired by a continuous supply of fairly small rock until an almost stable S-shaped equilibrium profile was reached. Among these are the breakwaters at Cherbourg in France, Plymouth in UK, Madras in India and Port Elliot in Australia.

In the 1960s, Priest *et al.* [1964] described a seaward profile, which is natural to the breakwater materials and the waves to which they are subjected. Experimental studies, often at a single water level, showed a stable, reshaped cross-section of an S-shape resulting in less wave action than on the initial steeper profile. It was concluded that a greater cross-sectional area was required for the breakwater forming the natural profile than for the conventional type cross-section. But, considering the possibility of using smaller stones than those indicated by conventional formulae, there might be instances where breakwaters with natural profile will compare favourably, in an economical sense, with those of conventional profiles.

In the late 1970s and early 1980s many researchers and engineers considered the idea of equilibrium slope and the importance of porosity or permeability, [Bruun and Johannesson, 1976], [Bruun, 1985]. The porosity of the armour, under-layers and core is a determining factor in the intensity of out and inflow which affects the stability of the structure. It was noted that the stability of rubble mound structures increases when "maturing". That is, the structure adjusts to wave attack and reaches an equilibrium profile or an S-shape with relatively small stone considering the wave climate.

1.2 Developments in Australia, mass-armoured breakwater

Australia played an important role in the development of berm breakwaters with many innovative structures being built in the 1970s and early 1980s. The earlier mentioned breakwater at Port Elliott in South Australia was one of the 19th century breakwaters constructed by dumping quarried rock into the sea and allowing it to take its own natural slope. At Grassy on King Island, Australia, a refinement of the dumped rock breakwater was achieved using available material from a nearby quarry and careful design, [Gourlay, 1996]. Here 95% of the material was less than 2 t and 5% was between 2 and 10 t. A core of quarry run material was pushed out to an offshore island in up to 18 m water depths, allowing waves to form it into a rocky beach before the large rock, 6-10 t, was placed on top of the reshaped profile for stabilisation.

During a cyclonic attack, the Rosslyn Bay breakwater in Queensland, Australia, suffered severe damage. At high tide the breakwater was heavily overtopped causing catastrophic failure. Material was displaced from the crest and deposited on the leeward slope, widening the profile while the crest was lowered by about 4 m. Still the reshaped breakwater showed a capability of protecting the harbour to some extent, [Foster *et al.*, 1978]. The breakwater was repaired by using commonly available rock sizes intermixed with modified concrete cubes with a grading that had the highest possible permeability, [Bremner *et al.*, 1987]. The construction procedures eliminated or reduced the use of a crane and simplified the construction by end-tipping with a minimum amount of trimming by dozer and backhoe. The design anticipated that natural wave action would reshape the seaward slope to the stable S-shape found in nature.

The experience from the Rosslyn Bay breakwater was used in the design of an offshore breakwater to protect a reclamation adjoining Townsville harbour in Queensland, [Bremner *et al.*, 1980], [Gourlay, 1996]. A shore-parallel offshore breakwater was built with crest level above high tide level. It was expected to fail or reshape under extreme wave conditions to form a submerged structure limiting the waves reaching the revetment protecting the reclamation. Extensive model testing showed that the design concept provided a considerable degree of safety against the design conditions being exceeded. Cost savings of the order of 40% were achieved over a conventional design, partly due to relative ease on construction not requiring large cranes.

The design of the Hay Point tug harbour in Queensland used the experience from these structures, [Bremner *et al.*, 1987]. Interpretation of preliminary quarry investigations and trial blasts in a nearby quarry assumed a maximum available rock size of 2-3 t. Further investigations, however, showed that it was possible to quarry armourstone of 3-7 t in large quantities. The development of design using these armourstone led to a definition of the mass-armoured breakwater that is designed and built in an initially unstable form, but with sufficient material provided to allow natural forces to modify its shape to a stable profile. Among the advantages of the mass-armoured breakwater is the use of natural rock in its available sizes.

1.3 Developments in Canada, modern berm breakwaters

Bill Baird and Kevin Hall from Canada initiated the design of what could now be called "modern berm breakwaters", [Hall *et al.*, 1983] and [Baird and Hall, 1984]. The idea was simple and effective with respect to design, construction and costs.

According to Baird and Hall [1984]:

> The basic principal involved in this concept is the use of locally available materials. It is established that the greater the thickness of the armour layer, the smaller the stones that are required to provide stable protection against wave action. Therefore, the thickness of the armour layer for a specific breakwater is determined by the gradation of the available armour stones and the incident wave climate. The final cross-section makes allowance for the practical considerations of breakwater construction. New concepts for breakwaters that have resulted from

the use of this alternative design procedure are described. Construction of these breakwaters in 1983-84 has demonstrated that significant cost savings are obtained.

In principle, the full quarry yield was divided into two classes: core and armourstone. The armourstone was used to create a homogeneous and permeable berm, including crest, and was constructed just by putting rock into the sea, as seen in Figure 1.1. This created a very steep seaward slope, often close to the angle of repose. The rock class was fairly small compared to a conventional stable structure and the first storms would partly reshape the berm into a more stable S-profile. Designs storms would give more reshaping until a large part or the whole berm was eroded and a stable S-profile was established. The easy quarrying (only two rock classes), easy construction and use of fairly small rock instead of large rock, or even concrete units, led to substantial cost savings.

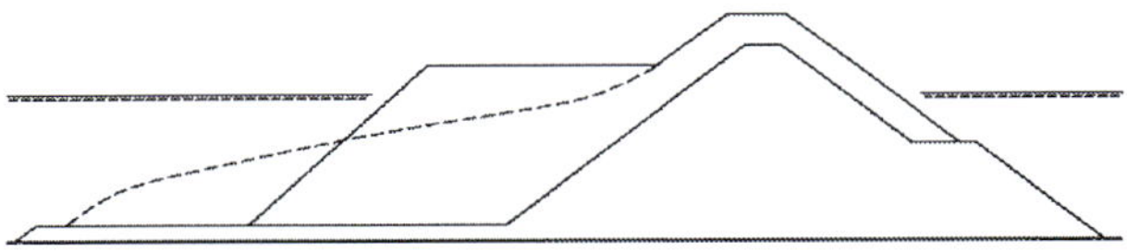

Figure 1.1. Principal sketch of first design of modern berm breakwaters in 1984.

Figure 1.2. Berm breakwater of Helguvik, Iceland. Designed in 1983 by Baird & Associates and constructed in 1986-1988 by Icelandic contractors.

By 1984, two berm breakwaters had already been constructed in Canada: Codroy in Newfoundland and North Bay in Ontario. The Helguvik breakwater in Iceland had also been designed by Baird & Associates, but was constructed a little later. Figure 1.2 shows the Helguvik breakwater more than twenty years after construction. Although the intention was to have a reshaping berm breakwater, the berm has hardly been reshaped during this period as quite some safety was used in the detailed design, mainly through the establishment of design wave height.

1.4 Contact between Canada and the Netherlands

The paper Baird and Hall [1984] was presented at the International Conference on Coastal Engineering in Houston. It was at the same conference where the first paper on new stability formulae of Van der Meer [Van der Meer and Pilarczyk, 1984], was presented, using the stability number $H_s/\Delta D_{n50}$ (H_s = significant wave height, Δ = relative mass density and D_{n50} = nominal diameter, see also Chapter 2). At that time, the research of Van der Meer was still in progress, with a focus for 1984 and 1985 on dynamically stable structures, like gravel or shingle and rock beaches.

The idea was to describe profile formation for dynamically stable structures, which indeed became possible (Van der Meer [1988-a]). There is a direct link to reshaping berm breakwaters, being the connection between full dynamically stable structures and statically stable conventional structures.

A week after the conference in Houston, Baird and Van der Meer met each other in Ottawa, discussing the berm breakwater concept and dynamic stability with profile formation. It led to restructuring of Van der Meer's research on rock slopes and gravel beaches, including some tests with berm profiles (Figure 1.3).

The focus of the research, however, was from gravel beaches towards reshaping berm breakwaters, not from static stability to more dynamic stability. Most of the berm profiles had stability numbers of $H_s/\Delta D_{n50}$ = 3.8-6.0, still far from statically stable reshaped berm breakwaters, where the stability number should not be larger than

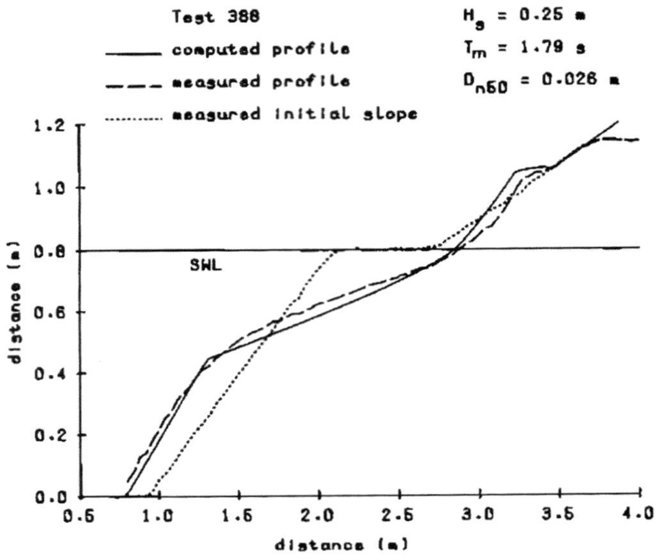

Figure 1.3. Berm-type profile in research on dynamic stability of [Van der Meer, 1988-a]. Test 388, including calculated profile by Breakwat; $H_s/\Delta D_{n50} = 5.9$.

$H_s/\Delta D_{n50} = 3.0$. Dynamically stable structures, like rock and gravel beaches, could well be described by the parameter H_oT_{om}, where T_{om} is a dimensionless mean wave period (see Chapter 2). Using this parameter means that a longer wave period will give a longer S-profile. It is this parameter H_oT_{om}, which still plays a role in description of the behaviour of dynamically stable structures, but less in berm breakwaters as the influence of the wave period is much smaller for berm breakwaters.

1.5 Developments in Iceland

During the preparation phase and model testing of the Helguvik breakwater in 1982 to 1983, undertaken by National Research Council (NRC) in Ottawa, the Icelandic Harbour Authority came to know about the berm breakwater design procedure. It was recognised that this design was very well suited for Icelandic conditions. At this time, the general opinion among engineers, geologists and contractors, was that it was difficult to quarry large armourstone from the Icelandic basalt. Several breakwater projects had been delayed due to the relatively large design wave height and lack of large armourstone for a conventional rubble mound design.

In 1983, two berm breakwater projects were initiated — a new breakwater at Bakkafjördur (see Section 9.5.2) and an overtopping protection of an existing pier at Hofsós. The existing pier at Hofsós showed too large and frequent overtopping (see Figure 1.4).

Figure 1.4. Overtopping during a storm at the pier of Hofsós before construction of the berm breakwater.

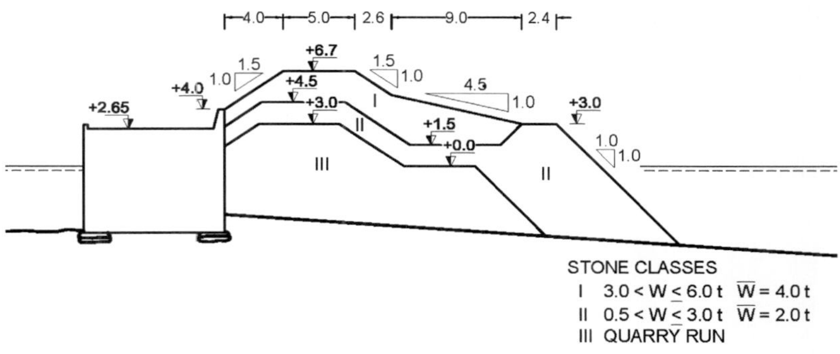

Figure 1.5. Cross-section of the design of the Hofsós berm breakwater from 1983 on the seaside of an existing pier for overtopping prevention.

Figure 1.6. The Hofsós breakwater in 2015 which was constructed in 1983. Photo by Indridi Einarsson.

The cross-section of the design is shown in Figure 1.5. Crest and upper slope use Class I rock, 3-6 t and $H_s/\Delta D_{n50} = 2.2$, with a narrow front berm and lower slope of Class II rock, 0.5-3 t and $H_s/\Delta D_{n50} = 3.0$. The breakwater at Hofsós was constructed in 1983 and since then, the berm has eroded, but the upper slope remains (see Figure 1.6).

Figure 1.7. The berm breakwater at Arnarstapi, Iceland, sheltering a small fishing harbour. Photo by Mats Wibe Lund.

The year after, in 1984, three more projects were initiated. At Arnarstapi an extension of an existing breakwater protecting a small fishing harbour was constructed. The berm structure at Arnarstapi was constructed from 1984-1985 and was rebuilt and extended by 45 m of conventional rubble mound in 2002 (see Figure 1.7).

Another new berm breakwater was constructed at Nordurfjördur (see Figure 1.8). With a stability number of $H_s/\Delta D_{n50} = 2.4$ on the trunk of the berm at Nordurfjördur, the berm has fully reshaped, while on the roundhead the berm holds its shape due to larger armourstone.

Figure 1.8. The berm breakwater at Nordurfjördur, Iceland, 30 years after construction (1984-1985). Photo by R. Kamsma.

A third and small berm type structure was constructed at Thorlákshöfn, preventing overtopping at the root of an existing breakwater.

During the subsequent years, several berm structures were constructed every year. In 1990, twelve berm breakwaters had been constructed, in 1995 — seventeen and in 2000 twenty-seven berm breakwaters had been constructed in Iceland, 50% of the known berm breakwaters worldwide, [Sigurdarson *et al.*, 2000]. All of these projects were designed and managed by a relatively small group at the Icelandic Harbour Authority, which in 1996 became the Icelandic Maritime

Administration and from 2013 is a part of the Icelandic Road and Coastal Administration.

In contrast to the reshaping approach presented by Baird and Hall, the Icelandic approach gradually developed into a design of a statically stable berm breakwater where only minor reshaping was acceptable. The reason was mainly due to the rock quality, as it was recognised that abrasion of the basaltic rock in Iceland could be high. To strengthen the berm, the size of armourstone in the two uppermost layers was increased to compensate for potential loss of weight, [Viggosson, 1990]. More stone classes were used compared to the original two stone class structure presented by Baird and Hall. While the stability number, $H_s/\Delta D_{n50}$, for the bulk of the rock berm was usually higher than 3.0, the stability number for the two uppermost layers was chosen as $1.2 < H_s/\Delta D_{n50} < 2.5$, so that only minor or limited reshaping was accepted under design wave conditions for the statically stable berm. These breakwaters showed less reshaping than the structures built with the original concept given by Baird and Hall [1984]. With less reshaping, it became possible to use smaller stones inside the berm, in areas that would not become directly exposed to the wave forces.

In the early phase of the development of the berm breakwater in Iceland, 3D physical model tests at scales of 1:45 and 1:60 were often used to test and refine the design. Both from testing and from some of the early berm projects, it became clear that the statically stable berm was excellent in reducing wave overtopping. During the eighties and nineties several concrete piers, which frequently experienced wave overtopping, were protected with a berm structure. Due to the high permeability, the berm practically swallowed up the incoming wave. Another feature recognised was the low reflection from these structures and especially from the breakwater roundheads. This increased the safety of vessels sailing through narrow harbour entrances and navigating in the vicinity of the berm structures.

Over the years, the design of the berm breakwaters developed in Iceland. The first structures were designed with a steep seaward slope as the reshaping berm breakwaters, with a natural angle of repose, often with the slope as steep as 1:1. This gradually changed to more gentle slopes, first to 1:1.3 and later to a much more stable slope of 1:1.5. The

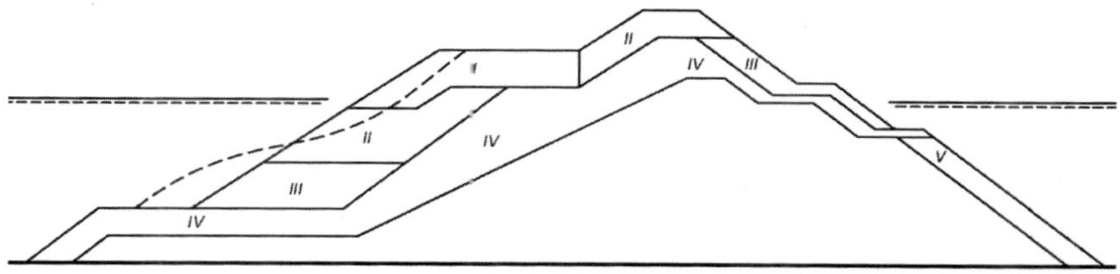

Figure 1.9. Principle cross-section of an Icelandic-type berm breakwater.

elevation of the berm did also develop from a low berm on the first structures to a higher berm. One of the initial reasons for this was that the excavators needed a working level for placing the top two layers on the berm of larger rock. The final result was a much less reshaping berm breakwater with more classes and large rock at the most attacked zone: the Icelandic-type berm breakwater (see Figure 1.9). It may be concluded that the Icelandic-type berm breakwater is significantly more engineered in contrast to the original berm or mass-armoured berm breakwaters and it includes more rock classes.

Construction methods were also taken into account and developed in cooperation with contractors through the projects. The experience from the first projects in Iceland was that using bulldozers to push rocks onto the berm generated too many fines that plug voids and diminish the berm permeability and wave energy dissipation. In later construction, armour placement was mainly performed by excavators of various sizes, which were able to place armourstone without the contamination of finer material. For short reach the excavators stood on the core but for longer reach, they were able to crawl on the smaller stone classes.

One of the basic ideas behind the berm breakwater concept was the utilisation of all size grades from the armourstone quarry. In Iceland all breakwater projects were executed with a dedicated quarry. The size of the large armourstone to be used for protection of the berm depended on the availability of large rock from the quarry. Therefore, a special emphasis was put on quarry investigations and quarry yield prediction was introduced as an integrated part of the Icelandic design procedure. At the same time, the blasting design developed to improve the yield of large armourstone. Lessons learned from one project were brought to the next by the project team and gradually it became possible to quarry large armourstone from the Icelandic basalt, but most often with a limited yield.

In the late nineties and early twenties, the concept of a statically stable berm design, based on utilisation of a dedicated local armourstone quarry and taking into account practical construction methods, came to be known as the Icelandic-type berm breakwater.

At the same time, several projects with a design wave height of about H_s = 7 to 7.5 m, were undertaken. This was a challenging task and demanded quarrying for armourstone heavier than 15 or 20 t. One of these breakwaters is the Sirevåg breakwater in Norway (see Figure 1.10). With a design wave height of H_s = 7.0 m and the largest stone Class I of 20-30 t, all quarried armourstone down to 1 t were utilised as well as all quarry run for the core of the breakwater.

Figure 1.10. Icelandic-type berm breakwater at Sirevåg, Norway, with 20-30 t class rock. Courtesy of the Norwegian Coastal Administration.

1.6 Berm breakwaters in international cooperation

The Workshop on berm breakwaters in Ottawa in 1987 [Berm breakwaters, 1987] was the first occasion in which berm breakwater knowledge was gathered. Later, European research work between 1990 and 1998 [MAST I, 1993] and [MAST II, 1997], gave more insight in various aspects of berm breakwaters and validated the idea that using available large rock on the right locations improved stability as well as it gave less reshaping.

A state of the art report on berm breakwaters was produced by PIANC MarCom Working Group 40 [PIANC, 2003], summarizing the research done so far and giving practical guidance for design and construction. This WG40-report also gave a classification of non-reshaping, partly reshaping and fully reshaping berm breakwaters, which describes the behaviour of different types of berm breakwaters. WG40 was not, however, able to derive accurate prediction formulae for recession of berm breakwaters.

More results of research on berm breakwaters became available after PIANC [2003]. This research has led to better understanding of the types of berm breakwaters present (mass-armoured with a homogeneous berm and Icelandic-type) and with a more precise classification (non-reshaping, partly reshaping and fully reshaping). Design guidance has been developed, using the existing formulae of Van der Meer [1988-a] for description of damage and an updated and more precise formula for recession of the berm.

1.7 Outline of the book

The first part of this book, Chapters 2–4, gives the more scientific background of berm breakwaters, including classification and development of design formulae for berm reshaping and wave overtopping. Chapter 5 mainly describes guidelines for the geometrical design of the cross-section. Chapters 6 and 7 give practical guidance on quarrying, project operation and construction. Chapter 8 uses all the information from the previous chapters to give direct design guidance for the different types and classes of berm breakwaters and for a design wave climate with wave heights between 3 m and 7 m. Finally, some examples of constructed berm breakwaters are described in Chapter 9.

Chapter 2

Classification and Types of Berm Breakwaters

2.1 Design wave climate, definitions and parameters

For planning, design and construction of a berm breakwater, the normal procedures have to be followed, as described in manuals like the Rock Manual [2007], or the Coastal Engineering Manual [2006]. It means defining project requirements, considering technical, cost, environmental and social matters, looking at rock quality and quantity, using design tools and considering construction methods. This book does not describe those aspects, as it has been specially focused on the berm breakwater structure, being one of many breakwater solutions. Work that is common for all kinds of breakwater solutions has not been described and can be found in the main manuals. This book may however update or replace sections 5.2.2.6 and 6.1.6 in the Rock Manual [2007], which are specifically devoted to berm breakwaters.

It is assumed that physical site conditions and data collection have been established and are available to be used in berm breakwater design. These are mainly the bathymetry, the hydraulic boundary conditions and geotechnical investigations data. Required design conditions are mainly on wave heights, wave periods, water depths, water levels and tidal variation for various return periods.

The conditions have to be described at the toe of the structure as these are the waves that will attack the structure and by using these conditions, the design methods will give a fair prediction of behaviour. In case the conditions change quickly in the last wave length before reaching the breakwater, it may be a good design decision to take the conditions half a wave length or more in front of the structure, instead at the toe. Such a decision will give a design with more safety, but may perhaps give a less accurate description of the behaviour of the structure.

The wave height to be used is the incident significant wave height H_s at the toe of the structure. Often the spectral wave height, $H_{m0} = 4(m_0)^{0.5}$, is taken as this wave height comes from wave climate studies. Another definition of significant wave height is the average of the highest third of the waves, $H_{1/3}$. This wave height $H_{1/3}$ is, in principle, the wave height that should be used in the Van der Meer formulae [Van der Meer, 1988-a]. In deep water, both definitions produce almost the same value, but situations in shallow water can lead to differences of 10-15%. In shallow water conditions, one may also consider the use of the $H_{2\%}$, especially if stability formulae are used.

In many cases, a foreshore is present on which waves can break and by which the significant wave height is reduced. There are models that can predict the reduction in energy from breaking of waves and thereby the accompanying wave height at the toe of the structure in a relatively simple way. The wave height must be calculated over the total spectrum including any long-wave energy present.

Based on the spectral significant wave height, it is reasonably simple to calculate a wave height distribution and accompanying significant wave height $H_{1/3}$ using the method of Battjes and Groenendijk [2000].

Various wave periods can be defined for a wave spectrum or wave record. Conventional wave periods are the peak period T_p (the period that gives the peak of the spectrum) and the average period T_m (calculated in the time domain from the wave record). The relationship T_p/T_m usually lies between 1.1 and 1.25. A wave period that is used more often in recent years, certainly in wave run-up and overtopping formulae, is the spectral period $T_{m-1,0}$ (= m_{-1}/m_0). This period gives more weight to the longer periods in the spectrum than an average period and, independent of the type of spectrum, gives similar wave run-up or overtopping for the same values of $T_{m-1,0}$ and the same wave heights. In this way, wave run-up and overtopping can be easily determined for double-peaked and "flattened" spectra, without the need for other difficult procedures. It is a wave period that has not been used a lot in berm breakwater design. For a single peaked spectrum, the ratio $T_p/T_{m-1,0}$ is close to 1.1. Any formulae can actually be rewritten by the use of this ratio and, in case of bi-modal or flattened spectra, the use of $T_{m-1,0}$ is then likely to give a more accurate prediction.

Wave steepness is defined as the ratio of wave height to wave length, $s_o = H_s/L_o$. Here L_o is the deep water wave length $L_o = 2\pi H_s/(gT^2)$. With use of T_p the steepness becomes s_{op}, with the mean period T_m it becomes s_{om} and finally, with the period $T_{m-1,0}$ it becomes $s_{om-1,0}$. The wave steepness tells something about the wave's history and characteristics. Generally a steepness of $s_{om} = 0.01$ indicates a typical swell sea and a steepness of $s_{om} = 0.04$ to 0.06 a typical wind sea. Swell seas will often be associated with long period waves. But wind seas may become seas with low wave steepness if the waves break on a gentle foreshore. During wave breaking, the wave period initially does not change much, but the wave height decreases. This leads to a lower wave steepness. A low wave steepness on relatively deep water means swell waves, but for depth-limited locations it often means broken waves on a (gentle) foreshore.

Stability formulae often include the stability number $H_s/\Delta D_{n50}$. Here Δ is the relative buoyant density, $\Delta = (\rho_r - \rho_w)/\rho_w$, where ρ_r is the mass density of the rock and ρ_w the mass density of the water. The nominal diameter $D_{n50} = (M_{50}/\rho_r)^{1/3}$, where M_{50} is the average mass of the stone class. Actually, D_{n50} is the "cubical" size of the rock with the average mass, regardless of the actual shape of the rock. Note that H_s is in the nominator and ΔD_{n50} in the denominator. An unambiguous notation would be $H_s/(\Delta D_{n50})$, but the notation $H_s/\Delta D_{n50}$ is used worldwide and will also be used in this book. Only when the stability number is used in combination with other parameters, the unambiguous notation will be used, for example in Equation 2.1

The cross-sectional shape or 2D-profile of dynamically stable structures like rock, gravel and shingle beaches, can be described by a dynamically stable profile [Van der Meer, 1988-a]. Such profiles change with the wave and water level conditions. The wave period has similar effects on the profile as the wave height, meaning that a longer wave period as well as a larger wave height would results in a "longer" profile. It is for this reason that Van der Meer [1988-a], introduced the dimensionless wave height - wave period parameter H_oT_o, which can be described by:

$$H_oT_o = H_s/(\Delta D_{n50})\cdot T\cdot(g/D_{n50})^{0.5} \qquad 2.1$$

With a mean period T_m the parameter becomes H_oT_{om} and with the peak period, H_oT_{op}. It is this parameter that also has been used in the past, like in PIANC [2003], to describe the recession of berm breakwaters.

Dynamically stable structures show some stability for a certain wave condition if a profile has been formed and this profile does not change as long as the wave conditions do not change. During each wave (wave breaking, wave run-up and wave run-down), it is possible that individual stones move up and down, but this does not affect the profile. This is of course not a good situation for a breakwater. A dynamically stable berm breakwater would mean that under severe wave conditions and after reshaping, some stones still move up and down the slope during individual wave action. This is different from the movement of rock during reshaping as in that situation, rocks move to a more stable position and then remain there.

Statically stable structures are stable under severe wave attack and only then may show a little movement of rock, leading to its so-called damage. Such damage, S_d, is related to the eroded area, A_e, around the water level, see Figure 2.1. This graph shows the original definition as given in Van der Meer [1988-a], with S (=S_d) as damage and A (= A_e) as erosion area. The definition of damage is:

$$S_d = A_e/D_{n50}^2 \qquad 2.2$$

For steep slopes (1:1.5 and 1:2) start of damage is given by $S_d = 2$ and filter or underlayer visible by $S_d = 8$. In the latter case about one rock layer has been eroded across the most attacked area. Individual stones of statically stable structures are fairly large compared to the wave height. Any single profile of the slope would show these individual stones, but by measuring one profile, it is impossible to detect an erosion area unless the damage is very large.

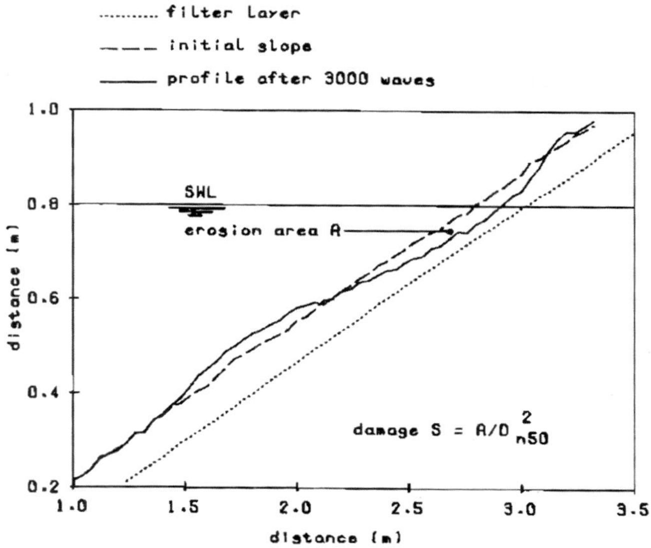

Figure 2.1. Definition of damage for a statically stable structure, [Van der Meer, 1988-a]

Hence, for statically stable structures, about 10 profiles are taken in any physical model test measurement and then averaged. This average profile smoothens the effect of individual stones and may detect values of S_d from $S_d = 0.5$ and larger. Figure 2.1 shows an average profile that was based on 9 individual profiles.

Dynamically stable structures can be given by a profile, statically stable structures by the damage S_d. But what about berm breakwaters? In many cases, the berm is not as stable as a statically stable structure and will reshape to some extent. PIANC [2003], used the recession of the berm, Rec, as a parameter to describe the behaviour of berm breakwaters.

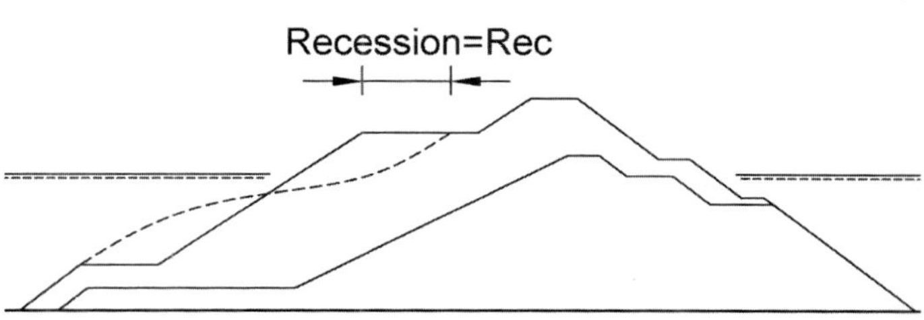

Figure 2.2. Principle idea of recession to describe the behaviour of a berm breakwater.

Figure 2.2 gives the principle idea of the recession parameter. A berm breakwater has large rock in the berm. Small recession, therefore, can only be detected if the average of many profiles is taken, according to the procedure for erosion area for statically stable structures. Only if the recession is quite large, as shown in Figure 2.2, three or even one profile would give a fair estimation of the recession. It can be concluded that large reshaping can be measured by a few profiles, but that for small reshaping the average of many more profiles is required. This has not been the case for many results presented in the past, which means that "small" recession has not always been presented with good accuracy.

Breakwaters should withstand design wave conditions and even more severe conditions. For very severe conditions like the 1000-years event or more, damage may be allowed, but failure of the structure should not occur. The ability to cope with very severe conditions is called resiliency. Some breakwater designs have better resiliency, or reserve capacity, than others. Also berm breakwaters may have better or worse resiliency with respect to wave attack. A fully reshaped berm breakwater does not have much reserve capacity left for further reshaping, as under very severe conditions the upper slope of the crest may be eroded and overtopping waves may start to erode the seaward side of the breakwater. Increased wave overtopping leads to damage of the seaward side, which may eventually lead to complete failure of the structure, where the crest height may reduce to heights around water level. On the other hand, a berm breakwater that has been designed to be almost statically stable under design conditions, still has the full berm to erode before overtopping will increase drastically.

It is clear that reshaping berm breakwaters have less resiliency than hardly or partly reshaping berm breakwaters and in this aspect, the resiliency or reserve capacity is a main difference between the different types of berm breakwaters.

The largest rock class or grading described in the Rock Manual [2007] is 10–15 t. This is the largest standard rock class that can be obtained from existing quarries that are used to produce standard rock classes only. The situation for berm breakwaters is often different, as dedicated quarries will be opened. For these dedicated quarries it may be possible to obtain rock over 20 t, even up to 35 t. There is no wide spread experience in winning these large rocks, requiring both the will and experience to quarry such rock. Note that only 2% or 4% of the total quarry yield is likely to be needed for the largest class of armourstone and this is one of the key points in succeeding to quarry such rock. More advice is given in Chapter 6, but in this book it is assumed that rock sizes up to 35 t may be available and can be used for design.

2.2 Classification of berm breakwaters by PIANC

The principle design of reshaping berm breakwaters, as described by [Baird and Hall, 1984], has been developed further into less reshaping and more stable berm breakwaters with more rock classes compared to only a small and a large class, core and rock. PIANC [2003] gave a classification of berm breakwaters as shown in Table 2.1.

Table 2.1. Classification of berm breakwaters by PIANC [2003].

Type of breakwater	$H_s/\Delta D_{n50}$	H_oT_{om}
Statically stable non-reshaped In this condition few stones are allowed to move, similar to the condition for a conventional rubble mound breakwater.	< 1.5–2	< 20–40
Statically stable reshaped In this condition the profile is allowed to reshape into a profile, which is stable and where the individual stones are also stable.	1.5–2.7	40–70
Dynamically stable reshaped In this condition the profile is reshaped into a stable profile, but the individual stones may move up and down the front slope.	> 2.7	> 70

The classification uses both the stability number $H_s/\Delta D_{n50}$ as well as the dimensionless wave height-wave period parameter H_oT_{om}. There is a large difference between the parameters as the stability number does not include any wave period effect and the H_oT_{om} parameter gives similar effect to wave height and wave period. PIANC [2003] does not conclude which parameter should be decisive for berm breakwaters, although formulae on recession of the berm have been given as function of H_oT_{om}.

The classes of statically non-reshaped and reshaped berm breakwaters both overlap the range of $H_s/\Delta D_{n50}$ = 1.5-2, although a different range in H_oT_{om} is given. This might suggest that a longer wave period results in more reshaping and a less stable profile. Proof of this, however, is not given.

Dynamically stable means that stones will be moving continuously under (severe) wave attack, which may lead to breaking of the stones and to longshore transport. Significant longshore transport affecting the stability of the structure should be avoided at all times. This is different from reshaping, where it is expected that displacement of stones will result in a more stable profile. One may actually conclude that dynamically stable (berm) breakwaters are not acceptable, as breakwaters during their life-time should be or become statically stable.

PIANC [2003] has given a classification of berm breakwaters that shows that there are different types of berm breakwaters with different structural behaviour. In that sense, the classification in Table 2.1 is very useful. But given the remarks above and developed insight in the stability of berm breakwaters, the classification can be updated.

2.3 New classification of berm breakwaters

Types of berm breakwaters

The homogeneous berm of a reshaping berm breakwater, as described by [Baird and Hall, 1984] has in many cases developed into a berm breakwater with more rock classes, where the largest rock class is present on top of the berm and partly on the front slope. In that sense it is possible to distinguish two types of berm breakwaters, the mass-armoured one and the Icelandic-type. The **mass-armoured berm breakwater**, with homogeneous berm, **MA**, is given in Figure 2.3.

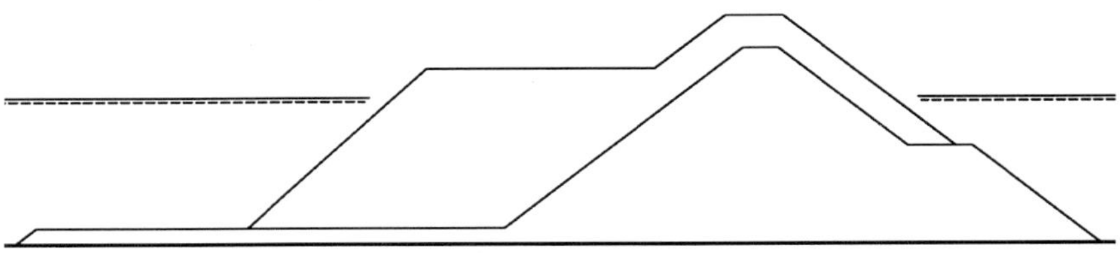

Figure 2.3. Mass-armoured berm breawater.

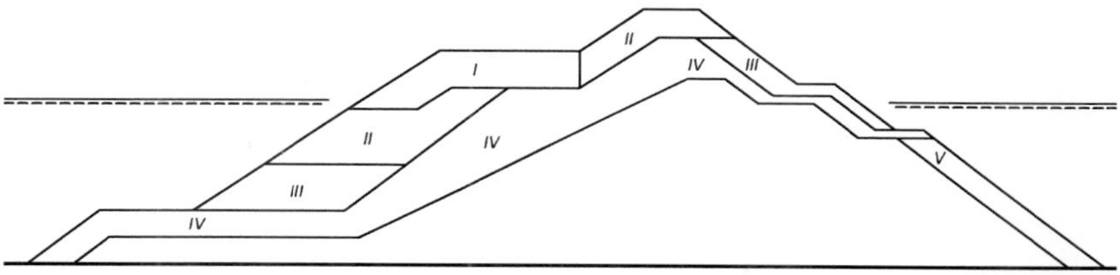

Figure 2.4. Icelandic-type berm breakwater

The **Icelandic-type berm breakwater**, with more rock classes and the largest rock class on the berm and partly on the seaward slope, **IC**, is given in Figure 2.4. Rock classes I to V range from the largest to the smallest stone class, often not overlapping.

The behaviour of both types will be very different if relatively small rock is used for the mass-armoured berm breakwater and very large rock for the Icelandic-type berm breakwater. The first one may fully reshape, where the second one may show static stability without significant reshaping. But it is also possible that similar rock classes are used and where both types may show partly reshaping. These types of breakwater do not always give similar behaviour, so the recession of the berm should be part of the classification.

Structural behaviour of berm breakwaters

Hardly reshaping	**HR**
Partly reshaping	**PR**
Fully reshaping	**FR**

Both the types of berm breakwaters and the different structural behaviour lead to a classification with four typical types of berm breakwaters:

Classification of types of berm breakwaters

Hardly reshaping Icelandic-type berm breakwater	**HR-IC**;	Figure 2.5
Partly reshaping Icelandic-type berm breakwater	**PR-IC**;	Figure 2.6
Partly reshaping mass-armoured berm breakwater	**PR-MA**;	Figure 2.7
Fully reshaping mass-armoured berm breakwater	**FR-MA**;	Figure 2.8

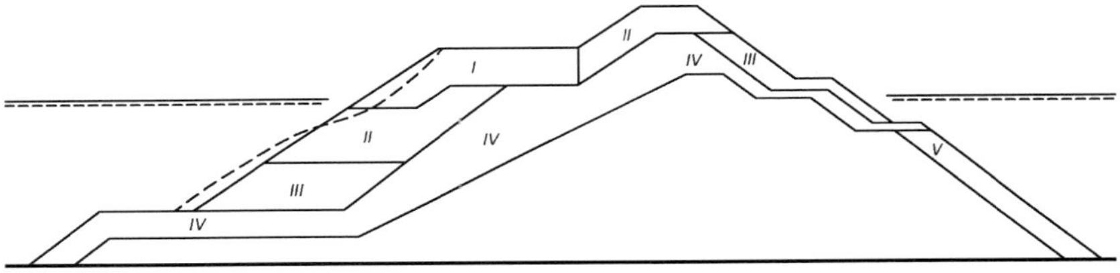

Figure 2.5. Hardly reshaping Icelandic-type berm breakwater.

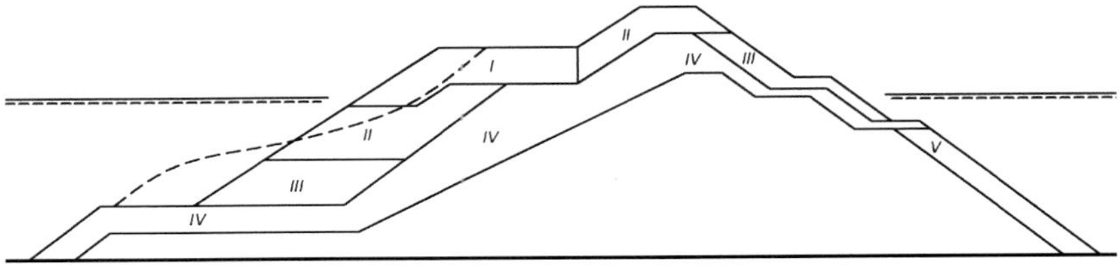

Figure 2.6. Partly reshaping Icelandic-type berm breakwater.

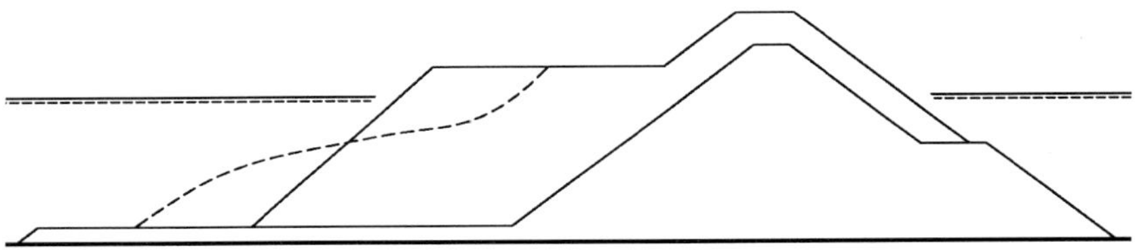

Figure 2.7. Partly reshaping mass-armoured berm breakwater.

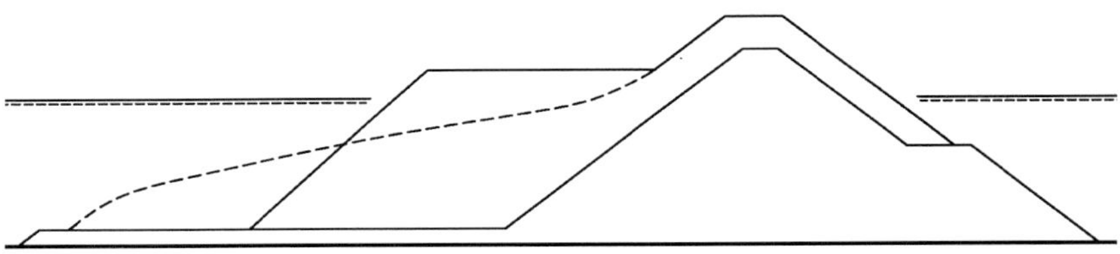

Figure 2.8. Fully reshaping mass-armoured berm breakwater.

The reshaping (mass-armoured) berm breakwater, **FR-MA**, Figure 2.8, has a large berm with mainly one rock class. The berm may be long and just above designed water level, but may also be quite high—similar to the crest height—and then with a narrower berm. As the berm will reshape, it is mainly the *volume or cross-section of the berm* that determines the design, not the width of the berm only. The seaward slope of the berm is often quite steep, up to the angle of repose of rock. The structure will be unstable when constructed, but should become statically stable after reshaping.

The behaviour is best described by the profile of the structure and only partly by recession. The construction should be easy without significant requirements and only a few rock classes have to be made, without very large sizes. A reshaping berm breakwater will be stable for overload conditions, but then not much resiliency will be left. Experience shows that this kind of berm breakwater may need some maintenance (adding berm rock) during the life-time of the structure.

The hardly reshaping Icelandic-type berm breakwater, **HR-IC** (Figure 2.5), will have a high berm and the total cross-section or volume will be less than a reshaping berm breakwater. The high berm has also a function as it will absorb the energy of large and long waves overtopping the berm. The hardly reshaping Icelandic-type of berm breakwaters need fairly large rock, but only in small quantities. Quarrying large rock using the right experience, shows that this will not significantly increase the cost of the structure in many cases.

Figure 2.9. The berm of the breakwater at Landeyjahofn, where specific care was taken to place the Class I stone to enhance stability.

The largest class of rock above the water line will be placed according to specifications that enhance stability. In fact, these stones can be placed in such a way that rocks on the outer layer are in contact with each other, with hardly any loose rocks. This specific placement, elaborated in Chapter 7 and shown as an example in Figure 2.9, delays the start of damage. It does not increase construction costs significantly as such big rock has to be placed individually.

The mass-armoured as well as the Icelandic-type berm breakwater may show partly reshaping when designed for it (**PR-MA**, Figure 2.7 and **PR-IC**, Figure 2.6). In this case, it is advised to have the berm level of the mass-armoured berm breakwater at the same level as the Icelandic-type. The ability to absorb wave energy with a relatively high berm is also good for the partly reshaping mass-armoured berm breakwater, as partly reshaping means that a significant part of the berm is left in place. There are, however, no placement specifications for the mass-armoured berm breakwater and the seaward slope still may be quite steep. If this is indeed the case, the first reshaping may be earlier than for an Icelandic-type berm breakwater, but after some reshaping they will show similar reshaping or stability.

The structural behaviour of berm breakwaters can be described by the recession, Rec, of the berm, as well as by the damage, S_d, if the reshaping is not significant. Certainly in case of not too large recession, like for hardly and partly reshaping berm breakwaters, the description in terms of damage is possible. The Van der Meer formulae [Van der Meer, 1988-a] show the influence of wave period on stability. For steep slopes and large permeability, a longer wave period often *increases* stability. This is further elaborated in Chapter 3. This is in contrast to using the parameter H_oT_{om}, which gives a large destabilizing effect for longer wave periods. It is for this reason, and also by proof of model test results, that the use of H_oT_{om} is more misleading than explaining and that it is better to base a classification primarily on the stability number $H_s/\Delta D_{n50}$.

The use of H_oT_{om} in earlier work like PIANC [2003] may be logical, however, as real dynamically stable profiles (rock and shingle beaches) have a strong correlation with this parameter, a longer wave period results in a longer profile. However, as mentioned before, all berm breakwaters should be considered as statically stable, even with a fully reshaped berm. And static stability cannot be described by H_oT_{om}.

Earlier work of Van der Meer [1990, 1994, 1995] mentions the definition of berm breakwaters as dynamically stable with stability numbers of $H_s/\Delta D_{n50}$ = 3-6. With hindsight one may now say that structures with stability numbers within this range are dynamically stable structures, but by no means (berm) breakwaters. This kind of structures will certainly be affected by longshore transport, like all dynamically stable structures. Berm breakwaters should be statically stable and the boundary from static to dynamic stability is close to $H_s/\Delta D_{n50} = 3.0$.

Finally, a breakwater may be subjected to all kinds of severe storms, from the 1-year storm up to for example a 1000-years storm. And breakwaters are often tested for the whole range of possible wave conditions, which leads to a range in stability numbers $H_s/\Delta D_{n50}$. In order to create a clear classification of berm breakwaters, it is proposed to use only the significant wave height, H_{sD}, for the *100-years return period* for the classifying stability number. In reality, smaller and larger values may be used during testing or for description of results, but the fixed return period gives a sharper classification.

Table 2.2 shows the new classification for berm breakwaters, including indicative values for the stability number, the damage and the recession. These values are given for a 100-years wave condition. For wave conditions with smaller return periods, the values will be smaller and consequently, for more severe wave conditions, like overload tests, the values may be larger. The values of Table 2.2 have been validated by tests, which have been described in Chapter 3.

Table 2.2. Classification of berm breakwaters, given for the 100-years condition.

Breakwater type	*Abbrevation*	$H_{sD}/\Delta D_{n50}$	S_d	Rec/D_{n50}
Hardly reshaping berm breakwater (Icelandic-type)	**HR-IC**	1.7-2.0	2-8	0.5-2
Partly reshaping Icelandic-type berm breakwater	**PR-IC**	2.0-2.5	10-20	1-5
Partly reshaping mass-armoured berm breakwater	**PR-MA**	2.0-2.5	10-20	1-5
Fully reshaping mass-armoured berm breakwater	**FR-MA**	2.5-3.0	--	3-10

By using the stability numbers of the various classes, it is possible to give an idea of the average mass of the rock class required for various wave climates. Table 2.3 gives an overall view of M_{50}-values, calculated using $\rho_r = 2700$ kg/m^3 and $\rho_w = 1030$ kg/m^3, for different stability numbers or wave heights.

Table 2.3. Required average mass M_{50} for various stability numbers and design wave heights H_{sD}.

	$H_{sD} = 3$ m	$H_{sD} = 5$ m	$H_{sD} = 7$ m
$H_{sD}/\Delta D_{n50} = 1.7$	$M_{50} = 3.5$ t	$M_{50} = 16.2$ t	$M_{50} = 44.3$ t
$H_{sD}/\Delta D_{n50} = 2.0$	$M_{50} = 2.1$ t	$M_{50} = 9.9$ t	$M_{50} = 27.2$ t
$H_{sD}/\Delta D_{n50} = 2.5$	$M_{50} = 1.1$ t	$M_{50} = 5.1$ t	$M_{50} = 13.9$ t
$H_{sD}/\Delta D_{n50} = 3.0$	$M_{50} = 0.6$ t	$M_{50} = 2.9$ t	$M_{50} = 8.1$ t

For a wave height of $H_{sD} = 3$ m one has the full choice of stability numbers or classes of berm breakwater as an average mass of 3.5 t would be easily available at many sites. This is already a little different for a wave height of 5 m, where for a stability number of $H_s/\Delta D_{n50} = 1.7$ an average mass is required of $M_{50} = 16.2$ t. Such an average mass requires a rock class of around 13-20 t, which is beyond the largest standard grading in the Rock Manual (2007). For design wave heights up to 7 m it has not yet been possible to design for a hardly reshaping berm breakwater with $H_{sD}/\Delta D_{n50} = 1.7$. For these wave heights the berm breakwater will be partly reshaping with a rock class of 20-35 t with an average mass of $M_{50} = 25$ t.

It is clear that the lower the stability number is, the more stable the structure will be, but also that there is a larger capability to cope with extremes. The resiliency from that point of view decreases with increasing stability number. One could say that a hardly reshaping Icelandic-type berm breakwater is very resilient to extremes, a partly reshaping mass-armoured or Icelandic-type will show good resiliency and a reshaping berm breakwater has marginal resiliency.

The classification above describes the expected *behaviour* of different types of berm breakwaters. It does not yet give design criteria. That will be done in Chapter 5, with worked examples in Chapter 8 for wave climates of 3 m, 5 m and 7 m wave height given in Table 2.3.

Chapter 3

Predicting Stability and Reshaping

3.1 Practical aspects

Most of the attention of berm breakwaters has been focused on the behaviour of the berm. This is more or less similar to conventional rubble mound structures, where the first design aspect is often the mass of the primary armour rock or concrete unit. The berm of a berm breakwater may reshape, but it may also be so stable that the behaviour can better be described as a conventional rock structure. The stability and/or reshaping of a berm breakwater is the focus of this chapter. Chapter 4 will treat other aspects on functional design (overtopping, transmission and reflection) and Chapter 5 will mainly describe the geometrical design of the cross-section and some other aspects on design.

Any loose material under wave attack in a two-dimensional situation, like with perpendicular wave attack in reality or in a physical wave flume, may give a stable profile. Sand dunes and gravel or shingle beaches and rock beaches give dynamically stable profiles, which change with every wave condition. Rock structures may be totally stable or may reshape to some extent. All these structures are stable. For dynamically stable structures, like shingle and sand beaches, the longshore transport governs design, not the cross-shore profile under wave action. For rock structures like breakwaters, longshore transport is not wanted and should be avoided.

For berm breakwaters any reshaping in principle is acceptable, as long as:

- the crest remains stable and overtopping does not damage the rear side too much;
- there is no longshore transport (statically stable after reshaping);
- there is no substantial breaking of rock.

This chapter does not describe the *design* of the berm of a berm breakwater, but the *behaviour* of berms under wave attack, mainly depending on type of berm breakwater and stability number. Chapter 5 gives practical design guidance, including aspects which govern decisions to be made when designing a berm breakwater. These aspects may be amongst others: resiliency or reserve capacity, wanted behaviour and expected maintenance.

Some of the analysis and guidance in this chapter has been published in conference proceedings, like Sigurdarson and Van der Meer [2011] on berm recession and like Sigurdarson and Van der Meer [2013] on berm recession as well as wave overtopping, reflection and transmission.

3.2 Existing prediction methods on static stability

The original Van der Meer formulae for statically stable rock slopes were published in Van der Meer [1988-a], but also in journal papers [Van der Meer, 1987-a] and [Van der Meer, 1988-b]. These formulae were also described in the first Rock Manual, [1991]. The new Rock Manual, [2007], however, treats a rewritten version of the original Van der Meer formulae, and added the so-called modified Van der Meer formulae for shallow water. It should be noted that these latter formulae were not modified by Van der Meer, but were based on the limited and confidential work of Van Gent *et al.*, [2003]. An elaboration on that work is given in Section 3.3, concluding that one should be very careful in applying those formulae.

Different from the Rock Manual [2007] only the original Van der Meer formulae will be applied here. The formulae can probably be applied to shallow water conditions where the significant wave height on the foreshore has reduced to a minimum of 50% of its original value on deep water.

The original formulae are given by:
For plunging waves:

$$H_s/\Delta D_{n50} = 6.2P^{0.18}(S_d/\sqrt{N_w})^{0.2}\xi_m^{-0.5} \tag{3.1}$$

and for surging waves:

$$H_s/\Delta D_{n50} = 1.0P^{-0.13}(S_d/\sqrt{N_w})^{0.2}\sqrt{cot\alpha}\,\xi_m^P \tag{3.2}$$

The transition of plunging (breaking) waves to surging (non-breaking) waves is given for:

$$\xi_{cr} = \left[6.2P^{0.31}\sqrt{tan\alpha}\right]^{\frac{1}{P+0.5}} \tag{3.3}$$

with:

H_s = significant wave height at the toe of the structure ($H_{1/3}$)
Δ = relative buoyant density: $\Delta = (\rho_r - \rho_w)/\rho_w$
ρ_r = mass density of rock
ρ_w = mass density of water
D_{n50} = nominal diameter: $D_{n50} = (M_{50}/\rho_r)^{1/3}$
M_{50} = average mass of the rock class
P = notional permeability factor
P = 0.1 = impermeable core beneath the armour layer
P = 0.5 = permeable core beneath the armour layer
P = 0.6 = homogeneous structure (only one rock class)
S_d = damage level (for slopes of 1:2 and 1:1.5: S = 2 start of damage; S = 8 underlayer visible)
N_w = number of waves in the considered sea state
ξ_m = breaker parameter: $\xi_m = tan\alpha/[2\pi H_s/(gT_m^2)]^{0.5}$
g = acceleration of gravity
T_m = mean period from time domain analysis
α = slope angle

Equation 3.1 applies for $\xi_{cr} < \xi_m$ and equation 3.2 for $\xi_{cr} > \xi_m$. For detailed description of the notional permeability factor P, applicable values for damage S_d, short sea states with $N_w < 1000$ and long seas states with $N_w > 7500$, as well as the use of $H_{2\%}$ for shallow water and the reliability of the formulae, one is referred to Van der Meer [1988-a],

[1988-b], [1987-a] or the Rock Manuals [1991] and [2007]. The easiest way to apply the formulae is to use the program Breakwat (commercial software from Deltares) as limitations with respect to the formulae are embedded in this program.

For berm breakwaters only one specific area of the stability formulae, equations 3.1 and 3.2, is of interest: the area with large permeability of the structure, say P = 0.5 - 0.6, together with a steep seaward slope of cotα ≤ 2. P = 0.6. This applies to a homogeneous structure (only one rock size) and could be used for mass-armoured berm breakwaters. The Icelandic-type berm breakwater also has large rock sizes, but if the largest rock size is considered as an armour layer, then other classes are a little smaller, but are still very permeable. A value of P = 0.55 would probably be applicable for Icelandic-type berm breakwaters.

The original tests of Van der Meer [1988-a] consisted of only one homogeneous slope of 1:2. Steeper slopes were not tested. For an armour layer with a permeable core (P = 0.5) slope angles of 1:3, 1:2 and 1:1.5 were tested. A slope of 1:1.5 is really the limit to be used in the formulae. Figure 3.1 shows the tendency of the formulae for these structures, including the original data, in a so-called $H_s/\Delta D_{n50}$ - ξ_m - graph. A larger breaker parameter gives a larger wave period.

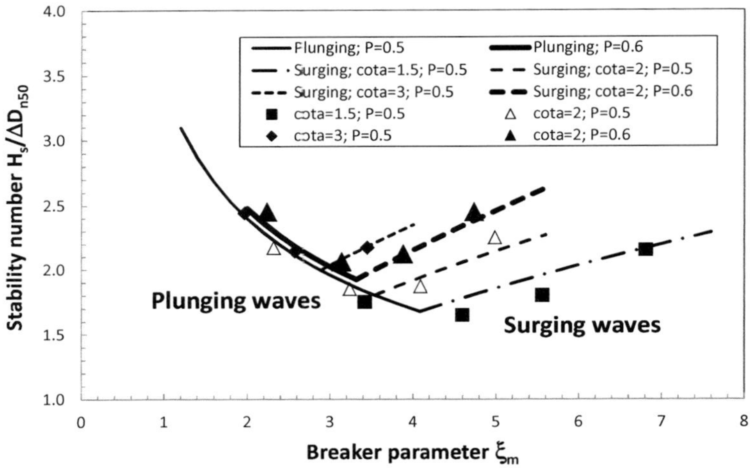

Figure 3.1. Stability results for a permeable core and a homogeneous structure (original data of Van der Meer [1988-a] with $S_d = 5$ and $N = 3000$) with Equations 3.1 and 3.2.

A homogeneous structure is more stable than an armoured structure even with a permeable core: compare the stability numbers for the triangles in Figure 3.1. Also a steeper structure is less stable than a gentler structure. The stability of a 1:2 slope of a homogeneous structure lies in between the results of a 1:2 and 1:3 structure with a permeable core.

It is clear from Figure 3.1 that there is a minimum of stability for a certain breaker parameter ξ_{cr} or mean wave period. Looking at hardly reshaping berm breakwaters, it is good to check this wave period for minimum stability. For a steep slope like 1:1.5 most of the wave conditions will be in the surging area, the right side of the figures. In this area the stability *increases* for increasing wave period. This behaviour may also be expected for berm breakwaters that are described as hardly reshaping and perhaps also for partly reshaping berm breakwaters.

The Van der Meer formulae are based on one test condition per test. After each test of 3000 waves the slope was reconstructed for the next test. In this way test results were not influenced by earlier test conditions. In reality, a structure will experience many storms of different attitudes. Also testing of breakwaters for design is often performed with a series of increasing sea states. The Van der Meer formulae can be applied to such situations by the cumulative damage method, which has been described in Van der Meer [1985-discussion] and later in the Rock Manual [2007] and which has also been implemented in Breakwat.

In fact the method is fairly easy. A first sea state, given by the significant wave height H_{s1}, or by $H_{2\%1}$, mean period T_{m1} and number of waves N_{w1}, gives a calculated damage level S_{d1}. A second sea state would be defined by H_{s2}, (or $H_{2\%2}$), T_{m2} and N_{w2}. The next calculation is to determine the number of waves N_{w12} that would be required for the second sea state to create the damage S_{d1} that was caused by the first sea state. Then the damage S_{d2} for the second sea state can be calculated, but now by applying $N_{w12} + N_{w2}$ as the number of waves. A third sea state would be calculated by repeating the same procedure.

Figure 3.2 gives an example calculation. The structure has a slope with $\cot\alpha = 1.5$, a notional permeability of $P = 0.55$ and a rock class with average mass $M_{50} = 10$ t and density $\rho_r = 2700$ kg/m^3. The mass density of the water is ρ_w = 1025 kg/m^3. Three consecutive sea states have been defined, each lasting for six hours:

Sea state 1; H_s=4 m; T_m=10 s; N_w=2160
Sea state 2; H_s=5 m; T_m=12 s; N_w=1800
Sea state 3; H_s=6 m; T_m=14 s; N_w=1543

For each of the sea states the S_d - N_w curve can be calculated using Equations 3.1 and 3.2. The damage development is a straight line for the first 1000 waves and then continues as a square root function $S_d \equiv N_w^{0.5}$. The damage after the first sea state can directly be calculated: S_{d1} = 2.24. This damage is reached for the second sea state for 585 waves. The damage for the second sea state is then calculated for N_w = 1800 + 585 = 2385 and amounts to S_{d2} = 5.92. This damage is reached with 738 waves for the third sea state and the final damage is then calculated with N_w = 1543 + 738 = 2281 and amounts to S_{d3} = 12.01. The damage for sea state 3 only would be S_d = 9.96, so the two first seas states increased the final damage by about 20%.

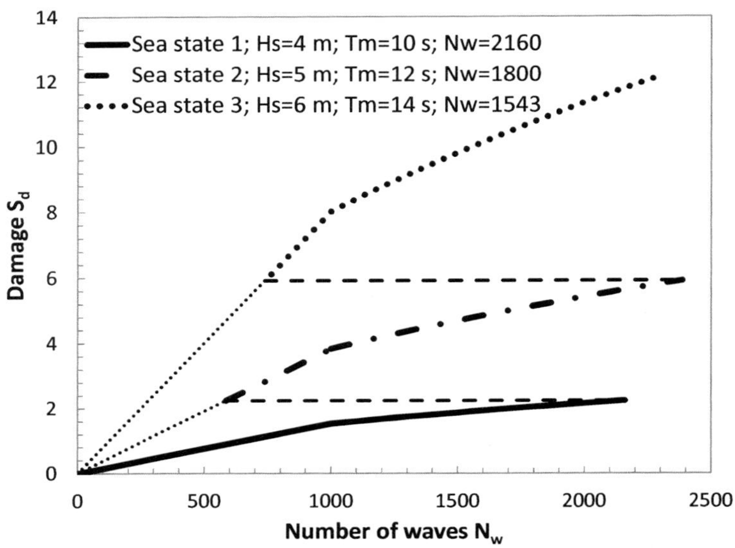

Figure 3.2. Cumulative method to calculate damage for various consecutive sea states.

3.3 The modified Van der Meer formulae in the Rock Manual

The Rock Manual [2007] gives the modified Van der Meer formulae for shallow water by equations 5.139 and 5.140. The re-fitting of the Van der Meer formulae was based on the work of Van Gent *et al.* [2003]. They applied a bulk analysis on their data, using the original Van der Meer formulae, but adopting different coefficients. Their work showed less stability in (very) shallow conditions and explanations for this were not given. Half of their tests were performed with bi-modal or double peaked spectra. For very shallow conditions it may well be that the trends in the Van der Meer formulae are no longer valid, like the relationship between damage and storm duration or number of waves, N_w: $S_d \equiv N_w^{0.5}$; or the relationship between damage and wave height: $S_d \equiv H_s^5$. This was not considered in the analysis by [Van Gent *et al.*, 2003]. The effect in the Rock Manual [2007] is that for rock slopes in shallow water one may end up with a (much) larger rock size than with the original Van der Meer formulae.

Another question is—to what conditions of "shallow water" would the original Van der Meer formulae be correct? It is true that most of the Van der Meer tests [1988-a] were performed for relatively deep water. One of the implications is then that the formulae are valid for Rayleigh-distributed wave heights. A limited number of tests were performed on a 1:30 foreshore with breaking wave conditions. These tests showed that in shallower conditions the distribution of wave heights is no longer according to a Rayleigh distribution and that less large wave heights occur. It means that in shallow water, the stability *increases* if the same significant wave height is present as in deep water. This is logical as less high wave heights are present in the shallow water case. For this reason, it has been proposed in Van der Meer [1988-a] to use the $H_{2\%}$ instead of the significant wave height (and adjust the coefficient in the formulae). The important trend in shallow water is that stability increases if waves start to break. Using the $H_{2\%}$-value in the Van der Meer formulae instead of the H_s leads for shallow water to a *smaller* required rock mass.

In very shallow water, however, other effects may play a role. If the original significant wave height reduces by breaking to 30% of its original value or even less, the wave steepness becomes very low and the shape of the waves may be very different from less shallow water situations, certainly with bi-modal spectra. Short waves or waves with high steepness cannot exist for these conditions. Secondly, the wave spectrum changes substantially with a tendency to have very long wave periods at the structure (based on $T_{m-1,0}$). For example the wave period may become four times larger than in deep water. The periods become so long that the breaker parameter is completely out of range of the original formula. Moreover, there are only very long periods, so it is likely that the influence of the period itself, which is dominant for the original formula, may be much less significant in the new situation. This will be of course speculation as long as data is not available.

Conditions such as these were mainly the subject of the research of Van Gent *et al.* [2003]. In their research, however, they never mentioned that the stability increases for first breaking of waves and they do not distinguish between first breaking and very large breaking—they just re-fit the formula with a new coefficient. That has led to the "modified Van der Meer formulae for shallow water", equations 5.139 and 5.140 in the Rock Manual [2007]. The conclusion is clear: this formula is not based on physical reasoning or a sound scientific analysis of processes and data and is actually fundamentally wrong. It is time for an updated errata to the Rock Manual [2007] stating this conclusion, as now many engineers and designers are using a fundamentally wrong formula.

3.4 Existing prediction methods on reshaping

Dynamically stable slopes, like gravel or shingle beaches and rock beaches were described by Van der Meer [1988-a]. A generic profile was schematised (see Figure 3.3), including heights and lengths for the crest and step (transition from the gentle slope below the waterline to a steep slope), some slope angles and also two slopes around the waterline described by a power curve. Most of the relationships were a function of the wave height—wave period stability number H_oT_{om}. The research covered the range of $H_s/\Delta D_{n50}$ = 3-256, where the value of 256 was reached by a significant wave height of 1.68 m on a 4.1 mm shingle beach in de large Delta flume of Deltares. The method to describe the profile was built in Breakwat and this module in Breakwat is an easy way to calculate dynamically stable profiles for all kinds of arbirtrary initial slopes. Application has been described in Van der Meer [1987-b] and Van der Meer and Koster [1988].

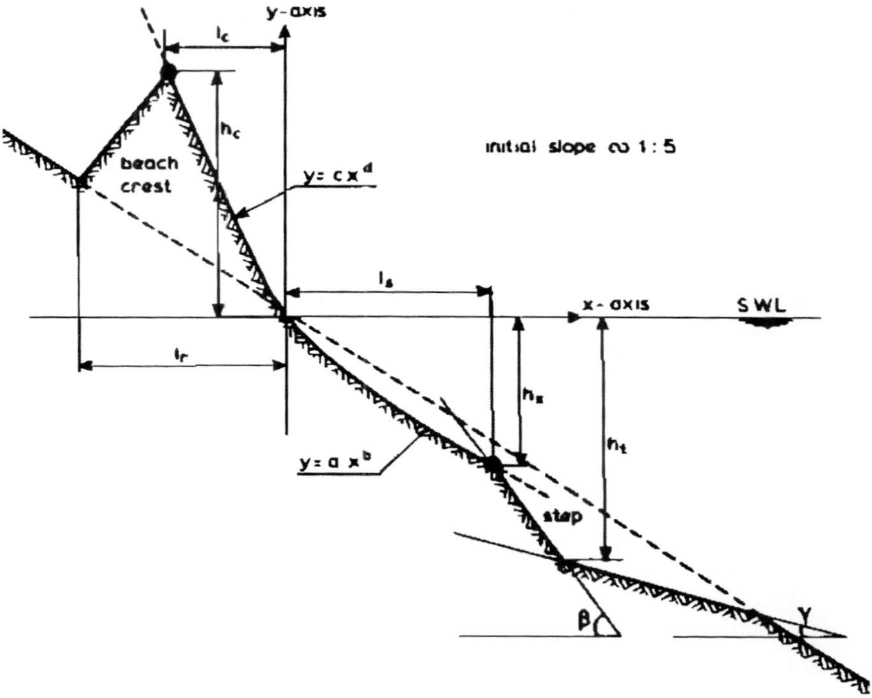

Figure 3.3. Schematised profile for dynamically stable structures like gravel, shingle and rock beaches [Van der Meer, 1988-a].

In Van der Meer [1992] the focus was on the lower end of stability numbers, from $H_s/\Delta D_{n50}$ = 3-6 and the paper was called: "Stability of the seaward slope of berm breakwaters". Breakwat was used to calculate various reshaped profiles, but all of them had stability numbers close to 3 or larger. In Chapter 2, it has been argued that (berm) breakwaters should be statically stable after reshaping and that stability numbers should be $H_s/\Delta D_{n50} \leq 3.0$ (for the 100-years condition, see Table 2.2). Only for overload conditions should stability numbers become larger than 3.

Although Breakwat is very well capable of describing the reshaped profile for "bermed" structures if $H_s/\Delta D_{n50} > 3$, the real minimum for Breakwat is about between $H_s/\Delta D_{n50} = 2.7 - 3$. And even then Breakwat may give a little too "long" profile, showing a little too much reshaping. An explanation for this can be found through the work of Moghim *et al.*, [2011], see further in this section.

There are no other methods that can describe profile reshaping from any arbitrary slope. But researchers have tried to model the profile for fully reshaping (**FR-MA**) mass-armoured berm breakwaters, like Hall and Kao [1991]. Their method has also been described in the Rock Manual [2007]—Section 5.2.2.6. The initial profile, however, was fixed to a lower slope of 1:1.25 and an upper slope of 1:3. It does not describe all variations that are possible in berm breakwater design, like different slopes below and above the berm, different berm heights, etc. Therefore, Hall and Kao's method has quite some limitations.

PIANC [2003] focused the attention on reshaping of the berm, and not on the description of the whole profile. A lot of data was gathered on reshaping (partly and fully reshaping) berm breakwaters and bermed structures, mainly based on Tørum [1998] and Tørum *et al.* [1999]. Some data were shown beyond static stability with stability numbers of $H_s/\Delta D_{n50}$ up to 4.5 with wave height–wave period numbers H_oT_{om} up to 160. The largest recession amounted to more than 30 stone diameters, which is far beyond static stability.

A recession formula was given, described by the wave height–wave period number H_oT_{om}. The scatter was very large and it was stated in PIANC [2003] that:

> Tørum (1998) could not find any explanations for the differences in the test results and attributed the differences to unknown differences in the test set-ups, test procedures, etc.

There are two main reasons for the large scatter found. The first is the use of H_oT_{om}. As described in Section 3.2, one may expect that a longer wave period does not result in a larger damage or recession. The use of H_oT_{om} for dynamically stable structures is good, but it should not be used for reshaped statically stable structures. The period gives a wrong influence on recession and this explains in part the difference. More will be validated on this in the next section.

A second reason for the scatter is the use of all data in one graph without distinguishing between the three types of berm breakwaters in the analysis and structures that are really dynamically stable under design conditions. Moreover, stability numbers larger than $H_s/\Delta D_{n50}$ = 3.0-3.5 will disturb the data if lower wave heights had been tested. A low wave height for a dynamically stable structure with a steep unstable slope 1:1 will show quite some reshaping. An Icelandic-type berm breakwater with similar stability number, but with a slope of 1:1.5 would hardly show any reshaping. Both data in the same graph will create quite some scatter.

The work of Moghim *et al.* [2011] focused on homogeneous (or mass-armoured) reshaping berm breakwaters. The range of stability numbers was $H_s/\Delta D_{n50}$ = 1.6-3.8, but the majority of the tests were in the range $H_s/\Delta D_{n50}$ = 3.0-3.8. All influences described in the analysis were given in this range. Actually they did not consider breakwaters, as a breakwater should be statically stable after reshaping, which means $H_s/\Delta D_{n50} \leq 3.0$. From that point of view, they considered dynamically stable bermed structures, which should not be confused with berm breakwaters. Such structures constructed in reality would during their life-time show large changes in profile due to longshore transport (like rock and shingle beaches).

But their work is interesting from another point of view. The range of stability numbers $H_s/\Delta D_{n50}$ = 3.0-3.8 lies perfectly between the statically stable berm breakwaters and the dynamically stable structures like rock and shingle beaches. For the berm breakwaters it is assumed that wave period may have little or no influence on berm reshaping (see Section 3.2). For really dynamically stable structures, the parameter H_oT_{om} is used to describe the profile and this parameter gives for wave height and wave period similar influence. Moghim *et al.* [2011] came to the conclusion that $H_oT_o^{0.5}$ described their test results best, giving some influence to the wave period, but less than to the wave height. This makes sense as the work of Moghim *et al.* lies between the statically stable and fully dynamically stable structures. It also gives the explanation why in the range of $H_s/\Delta D_{n50}$ = 2.7- 4 Breakwat may give slightly too long a profile. It should be concluded, however, that the work of Moghim *et al.* [2011] cannot be applied to much more stable berm breakwaters like the hardly or partly reshaping types.

Finally, the data of Lykke Andersen [2006] or Lykke Andersen and Burcharth [2010] should be mentioned. The first reference is the PhD-thesis of Lykke Andersen, the second reference is a summary journal paper. The 2006 work is used consequently as reference for ease of reading in this book, but one should also consider the journal paper, which may be easier available.

Lykke Andersen performed basic research on reshaping of berm breakwater profiles. Not all of the profiles can be classified as berm breakwater designs, as he varied berm width and berm level out of the usual "design range", but as more than 400 tests have been performed on reshaping, this data set is so large that it should be taken into account separately. He tested three rock classes, which were partly in the classifications as defined in Table 2.2: hardly reshaping, partly reshaping and fully reshaping. But also a significant part of the data had stability numbers $H_s/\Delta D_{n50} > 3.5$. His data will be re-analysed in depth in Section 3.7.

The final result of Lykke Andersen [2006)] was a formula to describe the recession of berm breakwaters over the full range of hardly reshaping into dynamically stable structures. This formula is a very extensive one, with a total of 14 parameters, 14 linear coefficients and 6 exponents. But in practice the basis of the formula has its limits. First of all, Icelandic-type berm breakwaters were not tested. The test sections were limited to mass-armoured structures, even for the hardly reshaping structures. Secondly, for the hardly reshaping and partly reshaping structures the berm was always quite low, lower than is used for Icelandic-type berm breakwaters. The effect of a high or higher berm was not investigated and, therefore, is not validated by his recession tests. Nevertheless, quite some tests fall into the category of berm breakwater.

Although the range of test results of Lykke Andersen [2006] has some limits, the formula was developed by considering also other research on berm breakwaters and the application of the formula is certainly wider than only on the basis of his own research. It is given here for completeness and possible application. It has been copied from section 4.2.14 of the reference, but parameters have been adjusted to the ones used in this book. For the full background of the formula one is referred to Lykke Andersen [2006].

The recession formula can be written as:

$$Rec/D_{n50} = f_{hb}\cdot[\{(1+c_1)\cdot h - c_1\cdot h_s\}/(h-h_b)\cdot f_{Nw}\cdot f_\beta\cdot f_{Ho}\cdot f_{skewness}\cdot f_{grading} + (\cot\alpha_d - 1.05)/(2D_{n50})\cdot(h_b-h)] \quad 3.4$$

where:

$$f_{hb} = 1.18\exp(-1.64d_b/H_{m0}) \quad \text{for } h_b/H_{m0} > 0.1$$

$$f_{hb} = 1 \quad \text{for } h_b/H_{m0} \le 0.1 \quad 3.5$$

$$c_1 = 1.2 \quad 3.6$$

$$f_\beta = \cos\beta \quad 3.7$$

$$f_{Nw} = (N_w/3000)^{-0.046H0+0.3} \quad \text{for } H_0 < 5$$

$$f_{Nw} = (N_w/3000)^{-0.07} \quad \text{for } H_0 \ge 5 \quad 3.8$$

$$h_s = 0.65H_{m0}\cdot s_{0,1}^{-0.3}\cdot f_N\cdot f_\beta \quad 3.9$$

$$T_0 = (g/D_{n50})^{0.5}\cdot T_{0,1} \quad 3.10$$

$$T_0^* = \{19.8\exp(-7.08/H_0)\cdot s_{0,1}^{-0.5} - 10.5\}/(0.05H_0) \quad 3.11$$

$$H_0 = H_{m0}/(\Delta D_{n50}) \quad 3.12$$

$$f_{H0} = 19.8\exp(-7.08/H_0)\cdot s_{0,1}^{-0.5} \quad \text{for } T_0 \ge T_0^*$$

$$f_{H0} = 0.05H_0T_0 + 10.5 \quad \text{for } T_0 < T_0^* \quad 3.13$$

$$f_{skewness} = \exp(1.5b_1^2) \quad 3.14$$

$$b_1 = 0.54Ur^{0.47} \quad 3.15$$

$$Ur = H_{m0}/\{2h(k\cdot h)^2\} = H_{m0}\cdot L_p^2/(8\pi^2\cdot h^3) \qquad 3.16$$

$$F_{grading} = 1 \qquad \text{for } f_g \leq 1.5$$

$$F_{grading} = 0.43 f_g + 0.355 \qquad \text{for } 1.5 < f_g < 2.5 \qquad 3.17$$

$$F_{grading} = 1.43 \qquad \text{for } f_g \geq 2.5.5$$

$$f_g = D_{n85}/D_{n15} \qquad 3.18$$

with:

Rec = recession of the berm of a berm breakwater

H_{m0} = significant wave height at the toe of the structure calculated from the spectrum

h = water depth at the toe of the structure

h_s = step height in the reshaped profile, see Figure 3.4

h_b = water depth above the berm, which is negative if the berm is elevated above SWL (note: in this book d_b will be used as the level of the berm with respect to DWL, the Design Water Level, and is positive if the berm is above DWL)

H_0 = stability number $H_{m0}/\Delta D_{n50}$ (note: in this book $H_s = H_{1/3}$ is used for stability, giving H_o)

α_d = structure slope of the berm of a berm breakwater

β = angle of wave attack with respect to the structure

$T_{0,1}$ = mean period based on frequency domain analysis (note: in the Van der Meer formulae the mean period from the time domain analysis is used, T_m)

$s_{0,1}$ = fictitious wave steepness for mean period = $2\pi H_{m0}/(gT_{0,1}^2)$

k = wave number = $2\pi/L_p$

L_p = peak wave length calculated from the linear dispersion relation

D_{n85} = nominal diameter, or equivalent cube size, $D_{n85} = (M_{85}/\rho_r)^{1/3}$

D_{n15} = nominal diameter, or equivalent cube size, $D_{n15} = (M_{15}/\rho_r)^{1/3}$

M_{85} = mass of particle for which 85% of the granular material is lighter

M_{15} = mass of particle for which 15% of the granular material is lighter

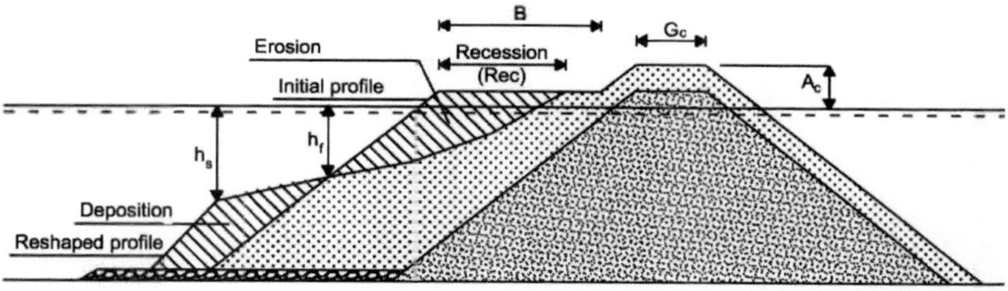

Figure 3.4. Definition of h_s in the recession formula of Lykke Andersen, [2006].

In some cases Equation 3.4 predicts negative recession. In these cases the recession should be zero. If a toe, which is wide enough to support the entire reshaped profile is present, the water depth above the toe should be used in the equations. If the toe is not wide enough to support the entire profile, then one has to use a water depth between the water depth above the toe and the water depth without the toe. This could be based on an estimation of the movement of the profile at the bottom and then use the mean water depth over the distance the profile moves [Lykke Andersen, 2006]. If the structure has no berm, the recession is merasured at the crest, hence $h_b = -R_c$ should be used in the fomula, as h_b is negative when the berm is above SWL.

3.5 Damage profile for a statically stable straight slope

The stability formulae 3.1 and 3.2 in Section 3.2 give a prediction of damage for a straight rock slope. Another interesting aspect of damage is where the erosion (damage) is located on the slope and where the displaced rock has been transported to. As for the definition of damage one has to consider the *average* profile, measured over many sections, as described in Section 2.1 and Figure 2.1. For statically stable straight slopes and even into dynamically stable straight structures (as a transition to the real dynamically stable structures given by Figure 3.3) one may use the stability formulae to predict the erosion area. Then one only needs those relationships for points on the straight slope where erosion turns into accretion and vice versa and a damage profile for a statically stable straight slope can be constructed.

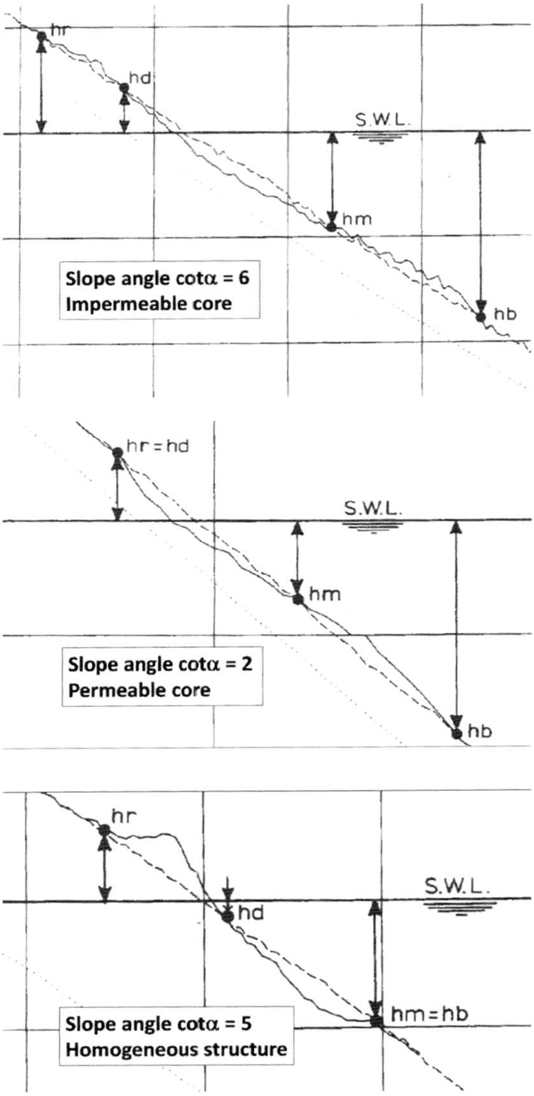

Figure 3.5. Statically stable damage profiles for straight rock slopes as found by [Van der Meer, 1988-a].

Figure 3.5 shows three examples of tests of Van der Meer [1988-a]. The profile can be schematised to an erosion area around SWL and accretion areas below SWL and/or above SWL. The transitions from erosion to accretion, etc. can be described by heights measured from SWL (see Figure 3.5). The heights are h_r, h_d, h_m and h_b respectively. If there is a beach crest, h_r will be larger than h_d and the accretion area will be between these two points. If there is no accretion below the erosion area, then $h_m = h_b$.

The relationships for the height parameters were based on the tests described by Van der Meer [1988-a] and have never been published internationally, for the main reason that they were already part of Breakwat (see Figure 3.6). Version 3.0 and higher of Breakwat, however, do not have this calculation tool anymore. For this reason the relationships have been summarised in Appendix A.

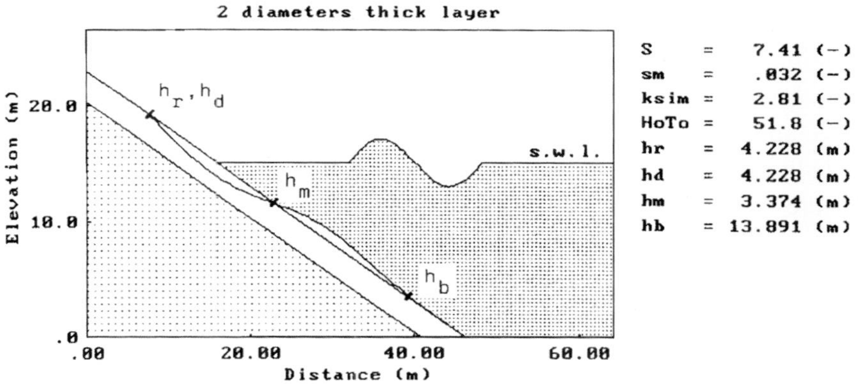

Figure 3.6. Statically stable damage profile for a straight rock slope as calculated by Breakwat [Van der Meer, 1992].

The assumption for the profile is a spline through the points given by the heights and with an erosion area (and one or two accretion areas), predicted by the stability formulae mentioned above. Of course, erosion area and accretion area(s) should be the same. The method is only applicable for straight slopes. The method can be used for statically stable structures, but also into dynamically stable structures until the procedure in Section 3.4, Figure 3.3 comes into play. If the berm of a hardly or partly reshaping berm breakwater is fairly high above SWL, the method may also give a reasonable estimation of the reshaped profile. The basic figure in the method, however, is the predicted damage (= erosion area). If that is quite good, the damage profile will be a reasonable prediction. The next section will look into that aspect.

3.6 Validation of damage prediction

Damage development is often studied by physical model tests with a certain number of consecutive sea states. Damage description for berm breakwaters is only useful if the reshaping is limited and actual damage values are more or less within the ranges used for statically stable structures. This means that only hardly reshaping and partly reshaping berm breakwaters can be considered.

A constraint is that damage should be measured accurately by the average of a number of profiles and the minimum number should be in the order of ten. Less profiles will lead to a lack of reliability, certainly if damage values are small. The damage, therefore, depends on the damage developed during previous sea states. This means that Equations 3.1 and 3.2 should be used with the method of cumulative damage, described in Figure 3.2. Breakwat was used for the calculation of damage in this section.

Sometimes data of real projects are available, but they may be restricted to some extent. Such data can only be treated in an anonymous way and a number of these projects will be described in this book. The projects are then simply called Project x.

Data of a real project became available and will be treated as Project 1. The design was an Icelandic-type berm breakwater with a design stability number of $H_s/\Delta D_{n50}$ = 1.7-1.8. The structure can be described as **HR-IC**: hardly reshaping Icelandic-type berm breakwater. Three tests have been performed on more or less the same cross-section, but attempts were made to improve the stability of the Class I rock.

It is hardly possible to place rock below the still water level (SWL) in a specific way, as visibility is very low under water. But from just below SWL upwards, this is well possible. And as large rock has to be placed individually, it does not take much longer to make an attempt in placing it in a more stable position. Section 7.4.1 gives more information on this aspect. For Project 1 rock lower than 1 m below SWL was dumped individually, but Class I rock above this level was individually placed and in such a way that it was hoped that it would increase stability. For the first two tests this was less successful as some damage to the dumped Class II rock below SWL also led to damage of Class I rock. The third test was very successful, showing no damage development, not even under overload conditions. The measures that were taken to increase the stability have been described in Section 7.4.1.

As there may be a difference between orderly placed rock and individually dumped rock, tests are distinguished in this book by the addition **OP** (orderly placed) or **D** (dumped). The tests for this project are then described as **HR-IC OP**.

Figure 3.7 gives the damage results for the three tests of Project 1. The tests started with lower wave heights than the 100-years design wave height and various sea states were repeated with slightly different water levels. Two overload conditions were tested, one with $H_s/\Delta D_{n50} = 2.0$ and a long wave period and one with a larger wave height with $H_s/\Delta D_{n50} = 2.4$, but with a much shorter wave period, giving a H_oT_{om}-value that was smaller than with the lower wave height.

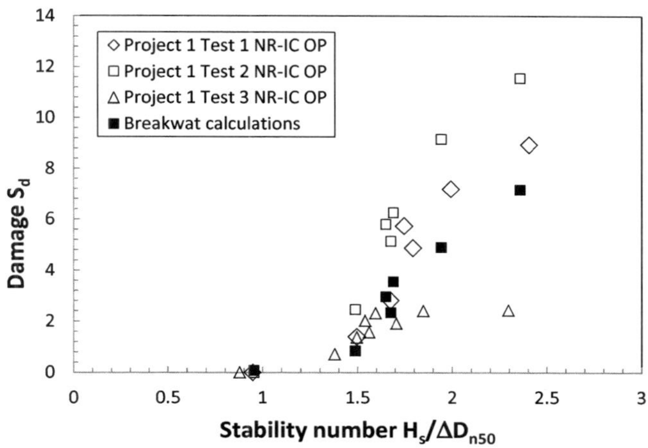

Figure 3.7. Model results and calculations of damage for project 1 HR-IC OP, using P = 0.55 and cotα = 2.

Damage occurred to both the Class I rock and the Class II rock below SWL. For damage calculations from the tests the total damaged area was taken and then S_d was calculated by using D_{n50} of the Class I rock.

"Start of damage" has been given as $S_d = 2$. Figure 3.7 shows that lower values up to $S_d = 0.5$ could be measured accurately and up to $S_d = 2$ the three tests show similar behaviour, mainly some rearrangement of rocks in the layer. Tests 1 and 2 show similar behaviour, where damage increases with increasing wave height. Damages are around $S_d = 2 - 6$ for design values of $H_s/\Delta D_{n50} = 1.7 - 1.8$ and they increase to $S_d = 9 - 12$ for the largest overload condition. Note that the largest wave height, with a lower H_oT_{om}-value than the previous sea state, gives a significant increase in damage. Wave height is determining the stability much more than the combination of wave height and wave period. Moreover, damage started at the Class II rock below SWL and this initiated unstable situations for the Class I rock just above this layer. The orderly placement did not work for these two tests.

The behaviour of test 3 was different. Small damage was developed up to the design conditions, which was mainly due to a small berm at the transition between Class I and Class II rock (see Section 7.4.1 for more details). But the damage did not increase for the overload situations, mainly due to the fact that the orderly placed rock remained stable.

Stability formulae 3.1 and 3.2 were used through Breakwat to calculate the cumulative damage. The seaward slope of the berm breakwater was 1:1.5, with a berm about one wave height above the design water level. Also a toe structure was present. The average slope from toe to crest would be about 1:2, which is more gentle than only the seaward slope of 1:1.5. As this was a statically stable hardly reshaping structure, it is better to use this average slope angle $\cot\alpha = 2$ in the formulae. An Icelandic berm breakwater has different classes of rock and although the berm consists of large rock only, it cannot be seen as a homogeneous berm. Therefore, a notional permeability factor of $P = 0.55$ was used instead of $P = 0.6$ (for a homogeneous structure). Results of the calculations are shown in Figure 3.7.

Up to design values the calculations are quite similar to tests 1 and 2. For the overload conditions, the calculated damage is lower than in these tests, but still in the same order. A reason for this under prediction may be that Class II rock contributes to the damage and this rock is smaller than the Class I rock. Test 3 under overload conditions is of course not modelled by the formulae, as they cannot predict stability for orderly placed rock. Overall, it can be concluded that for a hardly reshaping Icelandic-type berm breakwater as in Project 1 damage can be predicted quite well with the stability formulae for (straight) rock slopes, using $\cot\alpha = 2$ and $P = 0.55$ in the calculations. It should be noted that the initial slope should not be steeper than 1:1.5, as steeper slopes become less stable. Also stability formulae 3.1 and 3.2 should not be used with slope angles steeper than 1:1.5.

The same Project 1 considered also use of slightly smaller rock, but then as a mass-armoured breakwater, without any requirements for placement of rock, and with a steep (unstable) seaward slope of 1:1.1. The rock, however, was still fairly large and this design can be described as a partly reshaping mass-armoured berm breakwater, with dumped rock: **PR-MA D**. The design stability number was $H_s/\Delta D_{n50} = 2.3$.

In total, 14 sea states were used to check the behaviour of the berm, often repeating sea states with slightly different water levels. Table 3.1 shows the different sea states that were used and which are described by the stability number $H_s/\Delta D_{n50}$ and the wave height wave period number H_oT_{om}. Figure 3.8 gives the damage development of the whole test.

Table 3.1. Test conditions in Project 1 for the **PR-MA D** berm breakwater.

Sea state	$H_s/\Delta D_{n50}$	H_oT_{om}
Cond. 1	1.22	33.8
Cond. 2	1.97	62.4
Cond. 3	2.16	68.2
Cond. 4	2.28	71.3
Cond. 5	2.01	44.6
Cond. 6	2.64	64.6
Cond. 7	2.68	93.0

The first wave conditions reshaped the slope a little, due to the very steep and unstable seaward slope of 1:1.1. This led to damage values of

S_d = 12-14, which can be considered as more than one layer of rock removed (S_d = 8 gives "filter or underlayer visible" for a 1:1.5 conventional rock slope, which is equal to about one layer removed). The structure remained stable up to the design conditions (condition 4). Only under the overload the reshaping led to larger damage values, even up to S_d-values larger than 25, which can be considered as beyond limits for a conventional two-layer rock structure. But the whole berm was not reshaped, as the berm was designed wide enough and the structure showed only partly reshaping.

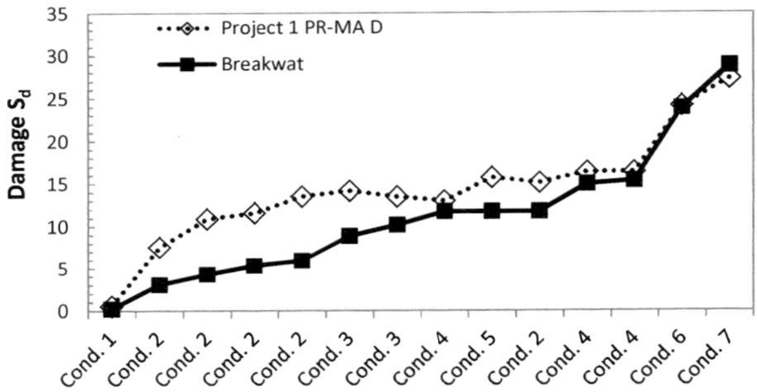

Figure 3.8. Damage progression during the test for project 1 **PR-MA D**. Calculations of damage by Breakwat using P = 0.6 and cotα = 1.5.

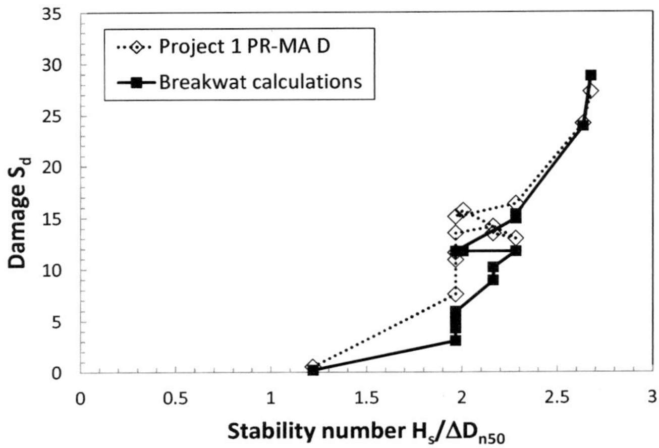

Figure 3.9. Damage as a function of the stability number $H_s/\Delta D_{n50}$ for project 1 PR-MA D. Calculations of damage by Breakwat using P = 0.6 and cotα = 1.5.

The results in Table 3.1 and Figure 3.8 can also be shown in another way, giving the influence of wave period on the damage parameter for partly reshaping berm breakwaters. Figure 3.9 gives the damage development as a function of the stability number $H_s/\Delta D_{n50}$. The middle of the graph shows that the wave height at some stages was reduced compared to an earlier state. In that case there was hardly any increase in damage. But for these conditions the wave period was much longer, giving a significant increase in H_oT_o-parameter. If the wave period would have substantial influence, damage should have increased significantly. This was not the case, neither in the Breakwat-calculations, nor in the test results.

Stability formulae 3.1 and 3.2 were not validated for unstable slopes steeper than 1:1.5. It is therefore strongly recommended to limit the use of the stability formulae to a slope angle not steeper than $\cot\alpha = 1.5$ for this kind of steep slopes and to accept that reshaping will probably give larger damage in the beginning. Where the berm is homogeneous with one rock class, the notional permeability should be $P = 0.6$. With these values Breakwat was used to calculate the cumulative damage during the test and results are also given in Figure 3.8 and Figure 3.9.

The calculations show indeed a less progressive start of damage than the steep slope in the model tests, but after some first damage and reshaping, the damage in the model tests slowed down and for design conditions the Breakwat prediction is quite close to the model results. This is even better for the overload conditions where prediction and model results are the same. This might be a coincidence, however, as the stability formulae were not validated for these large damage levels.

Also for this kind of berm breakwater the stability formulae gave a fair prediction of damage, using $\cot\alpha = 1.5$ and $P = 0.6$, although initial damage was under predicted. The prediction might have been better if the seaward slope of this berm breakwater would have been 1:1.5 instead of 1:1.

Damage values for fully reshaping berm breakwaters would be far beyond the range of application for the stability formulae 3.1 and 3.2. Therefore, damage prediction is only limited to hardly reshaping and partly reshaping berm breakwaters.

Recent research by Lykke Andersen *et al.* [2012] and Thomson *et al.* [2014] concentrated also on the use of the Van der Meer formulae (Equations 3.1 and 3.2) to predict damage for hardly reshaping berm breakwaters. From model tests Lykke Andersen *et al.* [2012] found that the Van der Meer formulae could be used to predict the damage for steep and hardly reshaping berm breakwaters (1:1.25), but they noted that the stability always followed the plunging formula (Equation 3.1) even in the surging regime (low wave steepness). From this, they concluded that the berm changes the type of wave breaking. Note that with a slope angle of $\cot\alpha = 1.25$ the range of application of the Van der Meer formulae was exceeded (maximum slope angle $\cot\alpha = 1.5$).

In Thomson *et al.* [2014] the stability of hardly reshaping berm breakwaters was investigated with relatively long wave periods. Based on these new model tests the statement by Lykke Andersen *et al.* [2012] above, that the stability follows the plunging formula also for low wave steepness, was proven incorrect if berm elevation and width was increased. A reduction factor for a berm was applied to the Van der Meer [1988-a] plunging formula. The factor was found to improve the estimation of the damage for some tests, but was insufficient to fully describe the damage. It was concluded that further tests with a larger range of berm widths, elevation and front slopes were therefore needed.

3.7 New method for recession of berm breakwaters

3.7.1 *Available data sets*

Although damage description for hardly reshaping or partly reshaping berm breakwaters may be a good tool for description of the behaviour of the berm, often the description has been limited to the recession of the berm only. The prediction of this recession is the subject of Section 3.7.

The reliability of recession measurements depends on two main aspects: the number of profiles used to measure reshaping and the definition of recession. As long as the reshaping is large, like for fully reshaping berm breakwaters, these aspects are less important as even one profile will show a large deviation of the initial profile. But it becomes more important if recession is small. For hardly and partly reshaping berm breakwaters, it is better to use an average of ten profiles or more. If laser scans are used, it is even possible to give the average profile over 50-100 points at every distance. Nice smoothed profiles are then obtained, which makes it much easier to determine more consistent values of recession.

Also the definition of recession plays a role. The principle sketch in Figure 3.10 is very clear, but reality is different. The damage or displacement of stones usually starts at around the still water level and then proceeds upward. For limited damage, as for the Icelandic-type berm breakwater, the damage might not have proceeded up to the top of the berm where the recession usually is measured. Another practical issue is that the recession on top of the berm is not easy to define. The front slope and top of the berm of the Icelandic-type berm breakwater is covered with large stones and when these are profiled, the profile rarely shows a sharp intersection between the front slope and top of the berm.

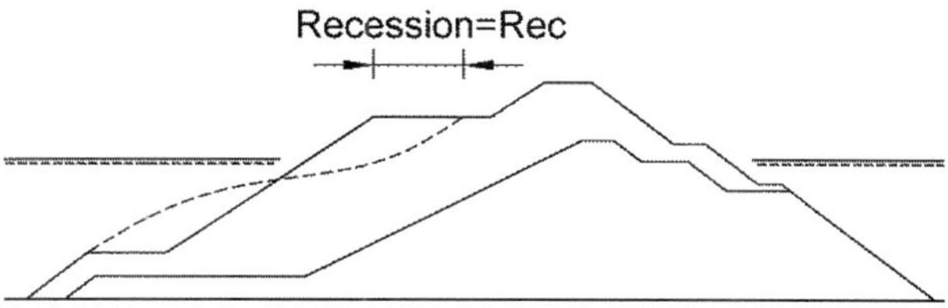

Figure 3.10. Principle sketch of definition of recession.

Therefore, it is necessary to define recession to take account of profile development on the full slope from top of the berm down to the water level, not only on top of the berm. The recession is taken as the horizontal difference between the as-built profile of the Class I armour and the profile recorded after the test, and between SWL and the crest of the berm (see Figure 3.11). The average recession distance, Rec, is the recession of the average profile, averaged between the water level and top of the berm.

Earlier work on recession may have been based on fewer profiles than the minimum considered and interpretations of the simplified definition in Figure 3.10 may have been different. At least the number of profiles measured should be considered if available data sets will be re-analysed.

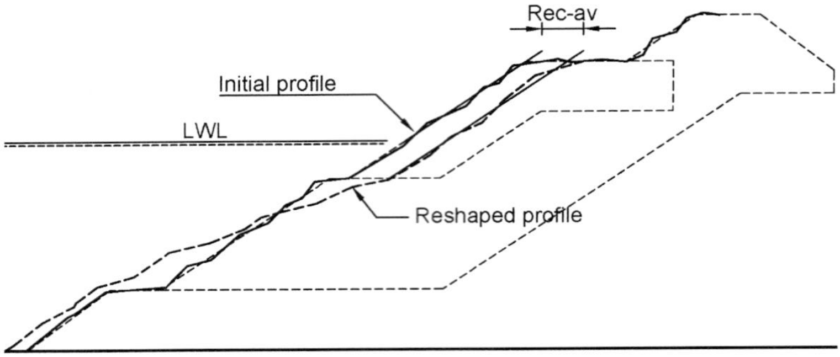

Figure 3.11. Modified definition of recession

A number of data sets on reshaping berm breakwaters were retrieved. A large number of them are public domain or were received from authors on request. A number of data sets were project-related and these data sets could only be used in an anonymous way, like "Project x". For each data set, first the type of berm breakwater was established on basis of the design value of $H_{sD}/\Delta D_{n50}$ (see also Table 2.2):

HR-IC	Hardly reshaping Icelandic-type berm breakwater	$H_{sD}/\Delta D_{n50}$ = 1.7-2.0
HR-MA	Hardly reshaping mass-armoured berm breakwater	$H_{sD}/\Delta D_{n50}$ = 1.7-2.0
PR-IC	Partly reshaping Icelandic-type berm breakwater	$H_{sD}/\Delta D_{n50}$ = 2.0-2.5
PR-MA	Partly reshaping mass-armoured berm breakwater	$H_{sD}/\Delta D_{n50}$ = 2.0-2.5
FR-MA	Full reshaping mass-armoured berm breakwater	$H_{sD}/\Delta D_{n50}$ = 2.5-3.0

Note that the hardly reshaping mass-armoured berm breakwater **HR-MA** has been added. This is probably not a type of breakwater that will be designed and constructed in reality, as it will need very large Class I rock everywhere in the berm. It will be much more economical to use the Icelandic-type of design, as very large rock will only be required on top of the berm and along the upper part of the seaward slope. The reason that it was added is that one part of the research of Lykke Andersen [2006] described test results of this kind of breakwater.

The significant wave height in the stability number for classification should be the 100-years wave height or, in scientific research, about 80% of the largest wave heights tested. This wave height is based on the waves measured in the time domain ($H_{1/3}$) and not on the spectral wave height H_{m0} and is according to the approach by Van der Meer [1988-a]. In many cases $H_{1/3}$ is quite similar to H_{m0}, but certainly at shallow conditions with fairly steep foreshores and small steepness, $H_{1/3}$ may become substantially larger (10%–15% is possible) than H_{m0}. One should also note that the calculation of $H_{1/3}$, and other related parameters like $H_{2\%}$, are not well described by the method of Battjes and Groenendijk [2000] if the foreshores are relatively steep (say steeper than 1:50) and the wave steepness quite low (say s_{op} smaller than 2%).

In order to get some insight in the geometry of the berm breakwater also some geometrical parameters were gathered, again related to the above mentioned significant "design wave height, H_{sD}". These are:

The upper slope	$\cot\alpha_u$
The lower slope	$\cot\alpha_d$
The berm width	B/H_{sD}
The berm level	d_b/H_{sD}
The crest freeboard	R_c/H_{sD}
Toe depth	h_t/H_{sD}

The berm level d_b, crest freeboard R_c and toe depth h_t are defined with respect to the still water level. The toe depth is the depth of the level to where most of the reshaped rock will be displaced. If a toe is quite small and reshaping is significant (as for fully reshaping berm breakwaters), then the level in front of this small toe should be taken and not at the crest of the toe. The idea is that the distance rock can fall down along the lower slope which has an influence on the recession of the berm (see Figure 3.20). One needs more recession to get a similar "S-profile" if the rock can be displaced to deeper parts. It is also expected that a relatively small toe depth will contribute to a more stable structure.

Table 3.2 gives the overall view of the data sets retrieved. There are five projects that had to be treated anonymously. Project 1, also described in Section 3.6, considered all three types of berm breakwaters: hardly reshaping, partly reshaping and full reshaping, although tests were performed at different laboratories. Projects 2–5 were all fully reshaping berm breakwaters, where Project 4 also considered some partly reshaping types.

Table 3.2. Overall view of data used for recession analysis.

Case	Type	Placement	Number of profiles	$\mathbf{H_{sD}/\Delta D_{n50}}$	Lower slope $\cot\alpha_d$	Upper slope $\cot\alpha_u$	Berm width B/H_{sD}	Berm level d_b/H_{sD}	Crest freeboard R_c/H_{sD}	Toe depth h_t/H_{sD}	Number of tests
Hardly reshaping											
Project 1	HR-IC	OP	10	**1.7**	1.5	1.5	1.5	0.9	1.6	2.4	24
L. Andersen Armour 1	HR-MA	D	M	**1.8**	1.25	1.25	0-4	0.2-0.3	0.8-1.2	2.4/3.1	19
Partly reshaping											
MAST II [1996]	PR-IC	D	5	**2.1**	1.1	1.5	3.2	0.9	2.2	1.9	10
Sveinbjörnsson [2008]	PR-IC	D	1	**2.1**	1.5	1.5	2.5	0.1-0.6	1.0-1.5	5.0	17
Myhra [2005	PR-IC	OP/ D	(6)	**2.0**	1.5/1.3	1.5	1.8	0.9	1.35	1.6	6/6
Lykke Andersen [2008]	PR-IC	D	M	**2.1**	1.5/1.3	1.5	1.5	0.6	1.3	2.5	13
Project 4	PR-IC	D	80	**2.2**	1.25	3.0	3.0	0.9	1.3	1.5	6
Keilisness	PR-IC	D	1	**2.1**	1.3	2.25	2.9	0.3	1.4	2.9	3
Myhra [2005]	PR-MA	OP/ D	(6)	**2.0**	1.5/1.3	1.5	0.9	1.2	1.45	1.6	6/6
Project 1	PR-MA	D	10	**2.3**	1.1	1.5	2.6	0.9/1.8	1.4	2.4	28
Lykke Andersen [2008]	PR-MA	D	M	**2.2**	1.5/1.3	1.5	1.5	0.6	1.3	2.6	8
L. Andersen Armour 2	PR-MA	D	M	**2.5**	1.25	1.25	2.5-4	0.35	0.8-1.5	2.6	57
Fully reshaping											
Project 1	FR-MA	D	3	**2.9**	1.5	1.5	3.0	0	1.0	2.4	4
Project 2	FR-MA	D	1	**2.9**	1.5	3.0	3.2	0.5	1.0	1.4	7
Project 3	FR-MA	D	5	**2.7**	1.5	1.5	1.7	0.5	1.1	1.8/1.4	3/8
Project 4	FR-MA	D	80	**2.8/3.1**	1.25	1.5/3.0	3.0/2.6	0.9/0.7	1.5	1.6/2.4	3/7
Project 5	FR-MA	D	3	**3.0**	1.33	1.33	2.6/3.6	0.6/1.1	0.6/1.1	4.0	12
MAST II [1996]	FR-MA	D	5	**2.9**	1.1	1.5	3.2	0.9	2.2	1.9	14
Moghim [2009]	FR-MA	D	9	**2.8**	1.3	1.5	2.0	1.0	1.7	1.5/5.1	7/5
Dynamically stable (not considered)											
Lykke Andersen Armour 3	MA	D	M	**3.5**	1.25	1.25	2.5-5	0.35	0.8-1.5	2.0-3.4	0
Moghim *et al.* [2011]	MA	D	3	**3.2**	1.25	1.25	3.4/5.2	0.2-0.9	1.4-1.8	2.8-3.2	0
Lissev [1993]	MA	D	1	**4.5**	1.25	1.5	3.0	0.4	1.3	3.6	0

Explanation of abbreviations in Table 3.2

HR-IC	Hardly reshaping Icelandic-type berm breakwater
HR-MA	Hardly reshaping mass-armoured berm breakwater (scientific test)
PR-IC	Partly reshaping Icelandic-type berm breakwater
PR-MA	Partly reshaping mass-armoured berm breakwater
FR-MA	Fully reshaping mass-armoured berm breakwater
FR-IC	Fully reshaping Icelandic-type berm breakwater (scientific test)
OP	Orderly placed rock
D	(Individually) dumped rock
H_{sD}	Estimated 100-years design wave height or second highest wave height in a research series
M	Multiple lines for profile measurements

The last column in Table 3.2 gives the number of tests that was considered for analysis. This number may be significantly different from the number of tests in the original research if part of those tests were outside the range of statically stable structures ($H_{sD}/\Delta D_{n50} \leq 3$). Three data sets are given which are fully outside the range of static stability.

A few of the data sets considered orderly placement **OP** instead of simply dumping **D** (placement without any specifications). For the partly reshaping structures Myhra [2005] found that there was hardly any difference in results. This was mainly due to the fact that the Class II rock below water always initiated damage, making the orderly placed upper Class I rock less stable. This was also the case for Project 1 for the hardly reshaping case (see Section 3.6). Only with very specific measures in placement as well as geometry of transition between Class I and Class II rock (see Section 7.4.1) was it possible to increase stability and to decrease recession.

All seaward slopes were between 1:1.1 and 1:1.5. Although the range seems small, it should be noted that a 1:1.1 slope is close to the natural angle of repose and will be less stable (at least initially) than a 1:1.5 slope. Upper slopes were steep, except for a few upper slopes with a 1:3 slope and one with 1:2.25.

In most cases more than one profile was measured. Sveinbjörnsson [2008] and the Keilisnes project, [Sigurdarson and Viggosson, 1994], are exceptions for partly reshaping berm breakwaters as they used only one profile to determine fairly small recession. Sveinbjörnsson [2008] his data set showed sometimes no recession (Rec = 0) for fairly large stability numbers, which is not in agreement with other data sets. Actually, with one profile it is not possible to measure recessions less than 0.5 D_{n50} in an accurate way. His data with Rec = 0 were not considered in further analysis. Also in Project 2 only one profile was used, but for a fully reshaping berm breakwater recession is always significant and one profile gives fairly accurate data.

Myhra [2005] measured six profiles, but he determined the recession for each of the individual profiles. This gave a large scatter. In the following analysis, the average of the six recession values is taken, but it should be noted that this is not the same as the recession determined by the average profile from six measurements.

There are of course variations in berm width and berm level. The trend for berm width is that the average value increases with increasing design stability number. The data of Lykke Andersen [2006] was basic research with variation in a significant range and should not be considered in this case to describe a general trend in berm width for design purposes. Berm levels, certainly for the hardly reshaping and partly reshaping berm breakwaters, are almost one significant design wave height above design water level for one part of the data ($d_b/H_{sD} = 0.9$), where another part has berm levels that are significantly lower ($d_b/H_{sD} = 0.1$-0.6).

There is large variation in toe depth. Some tests have been performed with deep water and no toe structure or with a toe structure deeper under water, others have a high toe level with h_t/H_{sD} around 1.6.

Lykke Andersen [2006] performed a basic study on berm breakwaters and excecuted 416 tests with results on recession of the berm. He varied rock size, berm width and level, water depth and of course wave height and wave period. The ranges of berm width and berm level are given in Table 3.2. A large range of berm widths was tested, often with berms wider than required for an optimum design of a berm breakwater. On the other hand, high berms were not tested. A number of berms with a level below swl were tested, but this must be considered as berm breakwater designs that are hardly possible for real practice and these tests have not been considered here. In all other tests, the berm level was just above the water level with $B/H_{sD} = 0.20\text{-}0.35$. Certainly hardly reshaping and partly reshaping berm breakwaters have often higher berms in order to increase stability, see also Table 3.2. A high berm with large rock dissipates the energy of an up-rushing wave better than a low berm.

The set-up of the research of Lykke Andersen [2006] consisted of similar tests for three rock classes, from large rock to fairly small rock. They were called Armour 1, 2 and 3. Looking at "design" stability numbers (for wave heights that are about 80% of the largest wave height in a test series) the three rock classes fall only partly in the classification of the three types of berm breakwaters (hardly, partly and fully reshaping). Figure 3.12 shows all test results as a function of the stability number, with the new classification of types of berm breakwaters.

Armour 1 was stable with small reshaping, comparable to Project 1 and a "design" stability number of $H_{sD}/\Delta D_{n50} = 1.8$. Armour 2 gives a "design" stability number of $H_{sD}/\Delta D_{n50} = 2.5$, which is just at the transition from partly to fully reshaping berm breakwaters. In Table 3.2 it has been considered as partly reshaping, but it is clear from that table that all other structures in that class have smaller design stability numbers.

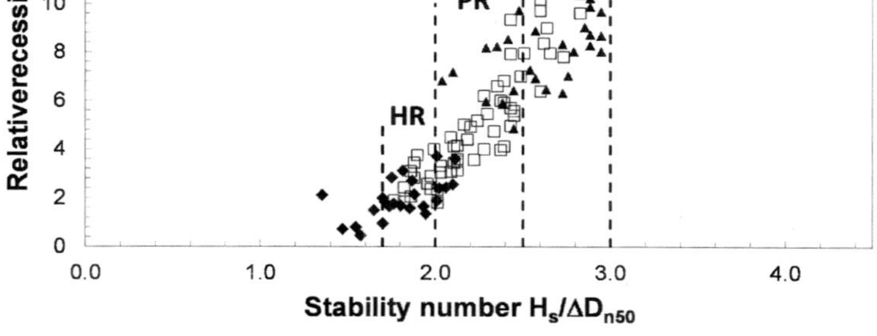

Figure 3.12. All recession results of Lykke Andersen [2006] with the new classification of types of berm breakwaters.

Most of the tests were performed with the small Armour 3, giving a "design" stability number of $H_{sD}/\Delta D_{n50}$ = 3.5, which is in the area of dynamic stability. These tests cannot be regarded as statically stable berm breakwaters and have not been considered further. In total, 36 tests on recession were performed for Armour 1, with some tests without recession and 57 tests for Armour 2, giving a total of 76 tests with recession measurements for further analysis.

A mass-armoured structure was tested in all cases by Lykke Andersen [2006], never an Icelandic-type of structure. Certainly for the hardly reshaping berm breakwater one would always decide on the Icelandic-type as this will limit the volume of the (very) large Class I rock drastically. Also all rock was dumped in place in the model tests and no specific requirements were taken for the Armour 1 rock to place them more stable. This is not a problem as often specific placement has no real effect. The seaward lower slope as well as the upper slope were held constant with $\cot\alpha_d = \cot\alpha_u = 1.25$. In fact this is a pity as for full reshaping berm breakwaters, often a very steep slope is taken, say 1:1.1, and for hardly reshaping and partly reshaping Icelandic berm breakwaters often a more stable slope 1:1.5. It means that even though the data set is large in size, it does not describe the effects of a fairly high berm, nor the influence of other slopes than 1:1.25.

Table 3.2 shows that in total 271 tests from 13 independent investigations were available for further analysis, with 76 tests from the data set of Lykke Andersen [2006].

3.7.2 *Development of new recession formula*

Figure 3.13 gives all recession data together, with the relative recession Rec/D_{n50} given as a function of the stability number $H_s/\Delta D_{n50}$. A fit, Equation 3.19, has also been given which will be discussed later. It looks like different types of berm breakwaters give a coherent picture with respect to recession, but still with a lot of scatter. Part of the scatter is caused by the data of Lykke Andersen [2006], where all data lie consistently above the fit. Further analysis will be first concentrated on the given design ranges of $H_s/\Delta D_{n50}$ for each classification and then each data set will be analysed in more depth.

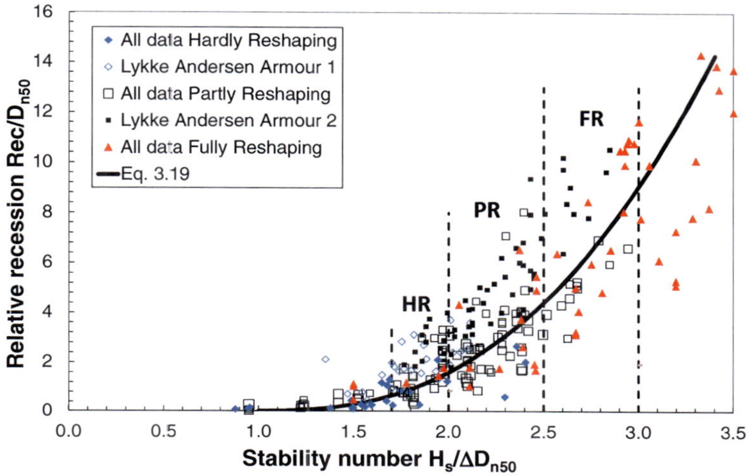

Figure 3.13. All recession data of the three types of berm breakwaters together.

Figure 3.14 to Figure 3.16 give again recession versus stability number, but now each figure gives one of the types of berm breakwaters. In these figures, the data of Lykke Andersen [2006] have been omitted. Those data will be treated later on. Focus of the analysis should first be on behaviour during design conditions and this area of $H_s/\Delta D_{n50}$ is marked by dashed lines in the figures. For lower wave heights, recession will be less and for overload conditions it will be more.

Figure 3.14 contains only data of Project 1 and the graph is quite similar to Figure 3.7, where the damage S_d was described. Recession was more or less similar for test 1 and 2, but much less for test 3, due to the success of orderly placement of the Class I rock. The design area with approximately the area with data points has been highlighted by the dashed line, with the centre of the design area in the middle.

Figure 3.15 gives some scatter, but the trend is very clear. There are significant outliers by the Keilisness data with at least $3D_{n50}$ more recession compared to the average trend. Although Keilisness was an Icelandic-type berm breakwater, the transition between Class I and Class II was close to the water level. This gave the effect that reshaping was governed mainly by the Class II rock and not by the larger Class I rock. The graph gives the recession using Class I rock as well as Class II rock. Note that these data are duplicated, but for different rock sizes. They will be analysed in depth later.

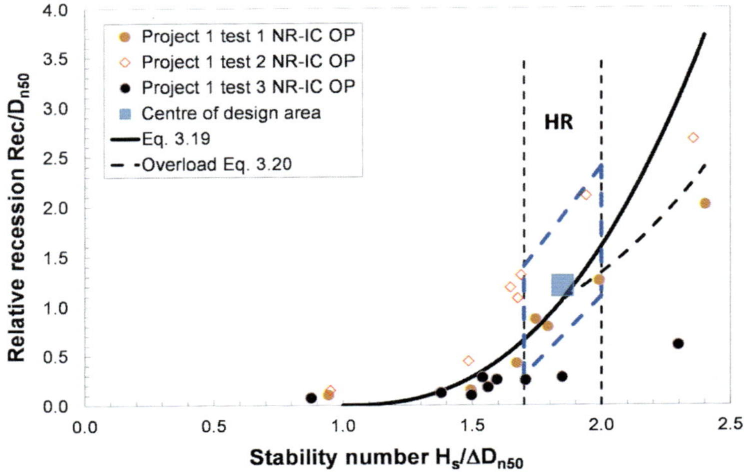

Figure 3.14. Recession for hardly reshaping Icelandic-type berm breakwaters **HR-IC**.

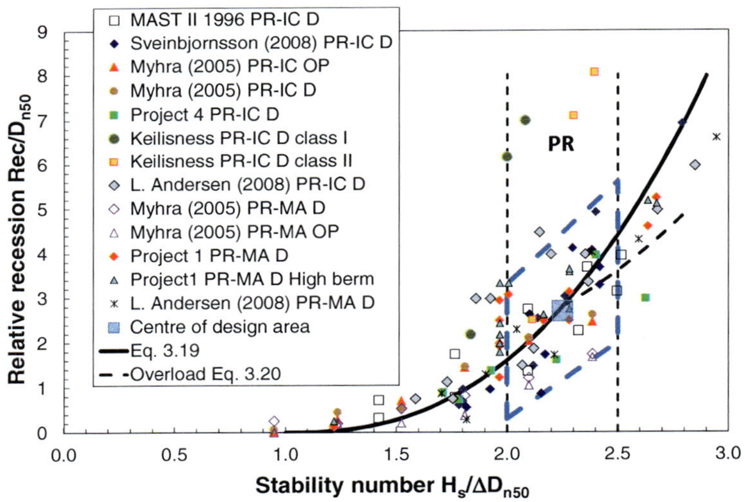

Figure 3.15. Recession for partly reshaping berm breakwaters **PR-IC** and **PR-MA**.

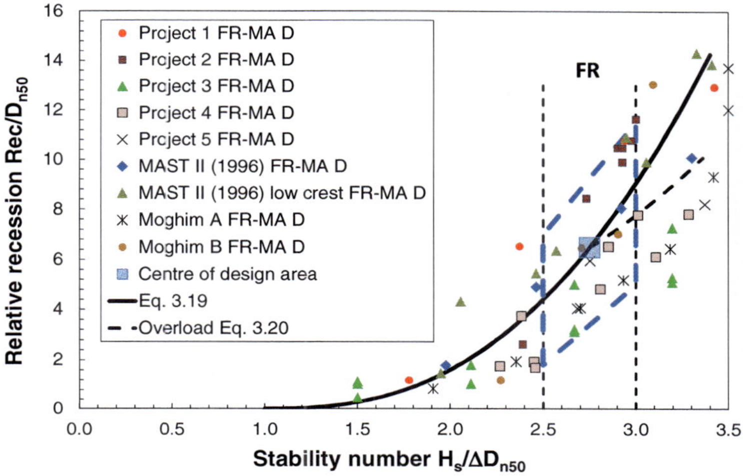

Figure 3.16. Recession for fully reshaping mass-armoured berm breakwaters **FR-MA**.

The full reshaping berm breakwaters in Figure 3.16 show also some scatter. In Figure 3.15 as well as Figure 3.16 some data sets lie consistently a little higher or lower than the fit. This means that there are more parameters that have influence on reshaping than only the stability number. These will be discussed further on.

Another observation is that the majority of the data points at the right side of the design areas in Figure 3.14 to Figure 3.16 are lower than the fit. This is a conclusion that often has been reached during model testing of berm breakwaters: most recession has taken place during pre-design conditions and design conditions. An overload condition will indeed increase recession, but not to the extent as before. The structure has already become quite stable and recession has slowed down. The dashed lines in the figures give a better average of the data in the overload situation than the continuous fit.

The wave height wave period number H_oT_o has often been used to describe the recession of berm breakwaters, see PIANC [2003]. Given the influence of the wave period for statically stable rock slopes, the influence should be much less or even opposite to the trend in this parameter (see also Section 3.2). In order to give an idea of the effect of

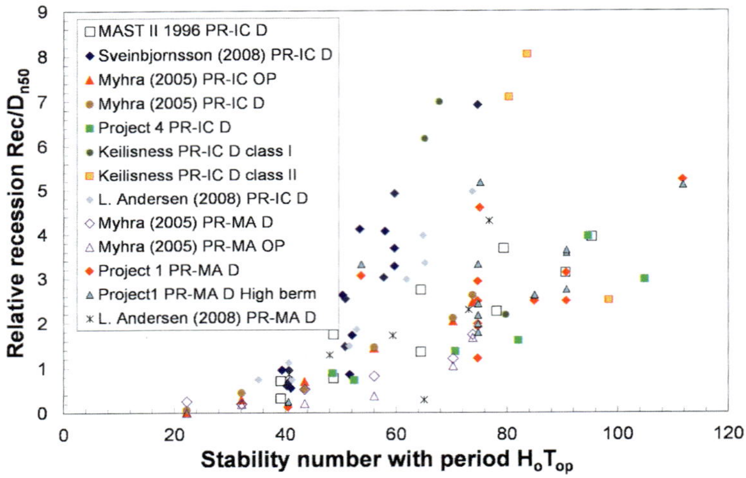

Figure 3.17. Recession for partly reshaping berm breakwaters, but now with the wave height wave period number H_oT_{op} instead of the stability number $H_s/\Delta D_{n50}$, showing very large scatter.

using H_oT_o instead of $H_s/\Delta D_{n50}$ Figure 3.15 has been repeated in Figure 3.17, but now by using H_oT_{op}. It is evident that the use of this parameter leads to a large scatter and this is certainly a reason why in PIANC [2003] no reliable fit was found for recession.

Stability tests are often performed for different sea states with increasing wave heights and wave periods. It is interesting to know the behaviour of the structure under design conditions and under overload conditions. It is for this reason that development of a new recession formula will be focussed on design conditions and not on all data, as the latter includes also results for small and larger wave heights than design conditions.

Figure 3.14 to Figure 3.16 were considered and for each given design range the centre of gravity of the data points was roughly established. Figure 3.18 gives these "centres of design areas", without the data. Then a fit was sought that would more or less give a trend through the three data points and that would start somewhere between $H_s/\Delta D_{n50}$ = 1–1.5.

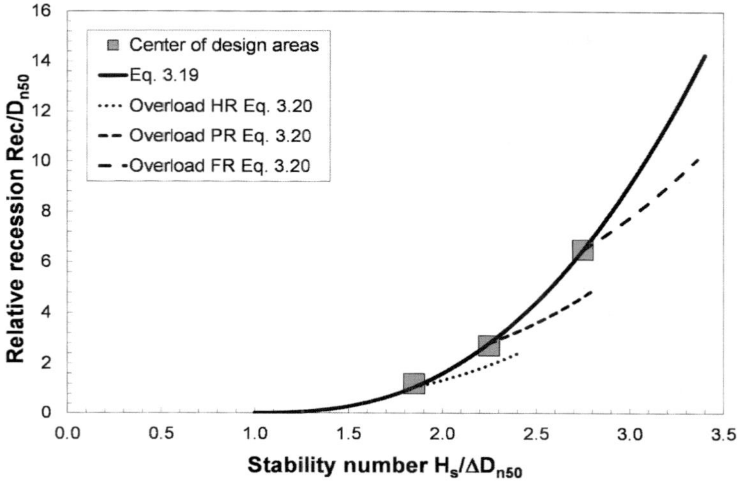

Figure 3.18. Centres of design areas for three types of berm breakwaters with continuous fit and predictions for overload situations.

The curve in Figure 3.18 is described by:

$$Rec/D_{n50} = 1.6\ (H_s/\Delta D_{n50} - 1.0)^{2.5} \qquad 3.19$$

Figure 3.13 gives all project-related data with Equation 3.19. It fits the whole range of data quite well, but still a check per type of berm breakwater is required. Figure 3.14 to Figure 3.16 give the data per type of berm breakwater, including the design area considered with the centre of it, and the new recession formula, Equation 3.19.

In all cases, the formula fits the design area as well as the part for lower wave heights. For the overload situations, in the Figure 3.14 to Figure 3.16 right of the dashed lines, there is a tendency that the formula over predicts the recession a little. It seems that each type of berm breakwater becomes quite stable after the design conditions and that overload conditions indeed increase recession, but not to the extent predicted by the formula.

Such a conclusion is difficult to reach if only the total picture with all data is analysed (see Figure 3.13), and should be taken into account during design of berm breakwaters. It means that Equation 3.19 can be used up to design conditions (100-years event), but that a smaller increase in recession can be expected during overload situations.

A practical method would be:

> Use Equation 3.19 up to design conditions;
> Calculate the recession with Equation 3.19 for overload conditions;
> Determine the final recession for these overload conditions by taking only *half of the calculated increase in recession in the previous step.* 3.20

For example, for design conditions, a recession of 5 m has been calculated and for the overload condition 8 m, using Equation 3.19, then the actual prediction of recession during overload would be 5 + (8-5)/2 = 6.5 m.

Equation 3.19 and the above method and figures give the general trend of recession as a function of the stability number only. Part of the scatter should be explained by considering other parameters that have an influence on recession. This is the subject of the next section.

3.7.3 *Influences on recession of berm breakwaters*

The stability number is the most significant parameter to describe the recession of berm breakwaters (see the previous section). But there are other influences on berm recession — some of them more or less hidden, and others that can be shown explicitly.

A more hidden parameter is cumulative recession. The basic research of Van der Meer [1998-a] on statically stable slopes (stability formulae 3.1 and 3.2) as well as on dynamically stable structures (Section 3.4 and Figure 3.3) was performed with only one wave height per test. Then, the structure was rebuilt and tested with other conditions. Also Moghim *et al.* [2011] followed this procedure for dynamically stable bermed structures. This method gives the damage or recession/reshaping for one wave condition only. The effect of a cycle of test conditions then has to be calculated by the method of cumulative damage (Section 3.2 and Figure 3.2).

For berm breakwaters, however, it is usual to follow a test sequence without rebuilding the structure. It means that previous test conditions have influence on the test results of a further test condition. Often the sequence is starting with low wave heights and increasing the wave height by small steps. But test series have also been performed where the next step was a lower wave height, but with a longer wave period. As the wave period has little or no influence, the result is testing with a lower wave height then before and no increase of recession. The data point in a recession graph then shifts to the left (lower stability number), but still with a fairly large recession (due to the previous test condition).

Some test series were performed with four or five steps with increasing wave heights. Other investigations had ten or more steps with changing wave heights as well as periods, or only the water level. Also changing water level may have some influence on recession, with similar or changing wave heights.

It is very difficult to include the way recession was established by the number of sub-tests, changing water levels and wave periods. These conditions will give some of the scatter in the recession graphs, but cannot be made more explicitly. Anyhow, all data sets considered measured the cumulative recession, which gives to some extent a similar procedure.

There are three geometrical parameters that may have substantial influence on berm recession. Some of them have also been described in the extensive formula of Lykke Andersen [2006], Equations 3.4 - 3.18. These are, with their dimensionless form:

- the lower slope $\cot\alpha_d$
- the berm level d_b/H_{sD}
- the toe depth h_t/H_{sD}

Lower slope

Conventional two-layer rock armour is not usually designed for slope angles steeper than 1:1.5. Steeper slope angles simply become too unstable and one needs very large rock to create a stable very steep conventional structure. Only with concrete units slope angles of 1:1.3 can be constructed, as concrete units may receive a substantial part of their stability from interlocking than from weight alone.

Practice (Table 3.2) shows that for berm breakwaters, a 1:1.5 slope is actually the most gentle slope considered! Very often, slopes of 1:1.25 or even 1:1.1 are considered, which are close to the natural angle of repose of the rock and these slopes are certainly not very stable. In some cases, like for the fully reshaping berm breakwaters, this is not a problem as large reshaping is expected. For hardly reshaping structures, however, a slope steeper than 1:1.5 will increase damage as well as recession. For partly and fully reshaping berm breakwaters the effect of a steep slope is shown in Figure 3.19.

The waves like to form similar S-profiles in Figure 3.19, but due to the steep slope in the left graph, the resulting recession will be larger.

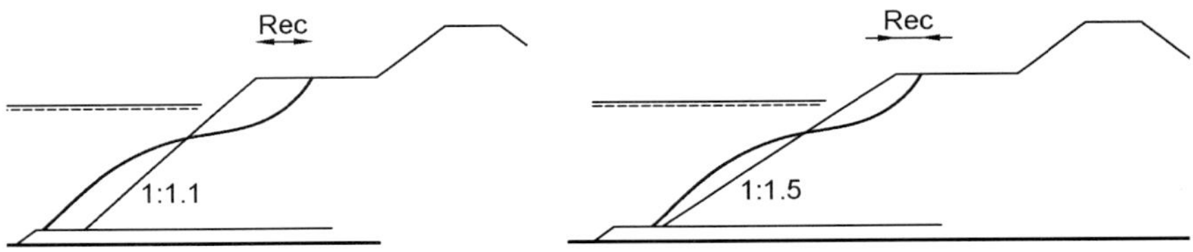

Figure 3.19. The influence of slope angle on recession for reshaping berm breakwaters.

Berm level

A low berm may give larger reshaping, as the waves need enough rock to create a stable profile and if this rock is not available above the water level, the waves will take it from the lower berm, creating a larger recession. A (very) low berm is sometimes considered for fully reshaping structures and will then show a larger recession. For hardly and partly reshaping berm breakwaters, the berm is generally well above the water level, as the berm acts as an area where up-rushing waves can dissipate their energy and this works better with a higher berm. Some datasets had fairly low berm levels and this could have effect on the recession.

As long as the berm is high enough above the water level, the energy dissipation of up-rushing waves will work. But an even higher berm will not have extra influence. This would mean that if the berm is higher than a certain threshold, it will work quite well; and if it is under this threshold the recession will increase a little.

Toe depth

The water depth in front of a berm breakwater may have some influence on recession, due to a different wave height distribution than the Rayleigh distribution. But knowledge about the wave height distributions is not available in the data sets.

The toe depth also gives a depth influence, but its effect is different than for the water depth. A water depth may change wave conditions, but the toe depth may have a direct influence on the reshaped profile, more or less regardless of the wave height. Figure 3.20 shows the effect. The toe depth is defined as the water depth above the lowest part in the built profile where eroded rock can go to. As said above, waves like to form a similar profile and if eroded rock can fall deeper, it needs more erosion to come to a similar "length" of the eroded profile. A small toe depth may drastically reduce the recession.

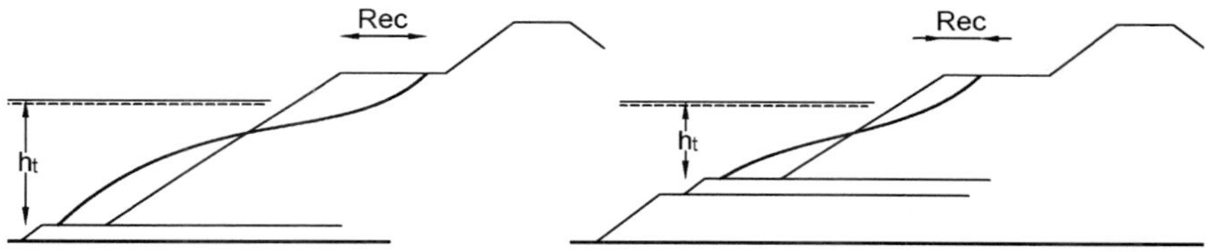

Figure 3.20. The influence of toe depth on recession for reshaping berm breakwaters.

The influence of slope angle, berm level and toe depth will be analysed in a comparative way as it is very difficult to extract enough specific results from the data on only one parameter. The influence may also differ depending on the classification of berm breakwaters: partly, hardly or fully reshaping. Based on the geometrical data in the data sets (Table 3.2), the location of the data points in the recession graph compared to the recession formula and the classification of the berm breakwater, a kind of "average" geometry has been assumed (see Table 3.3). Deviating from this geometry may then lead to a negative score (if assumed less stable), but also to a positive score. A structure parameter similar to the average would get a neutral score. Scores could be --, -, 0, + and ++.

Table 3.3. Positive and negative scores on berm reshaping. The second column gives the average condition with score 0.

Hardly reshaping berm breakwaters ***HR***			
lower slope	$\cot\alpha_d = 1.5$	$\cot\alpha_d < 1.5$	score: -
berm level	$d_b/H_{sD} \geq 0.6$	$d_b/H_{sD} < 0.6$	score: -
toe depth	h_t/H_{sD}	no influence, hardly any reshaping	
Partly reshaping berm breakwaters ***PR***			
lower slope	$\cot\alpha_d = 1.5$	$1.33 < \cot\alpha_d < 1.5$	score: -
		$\cot\alpha_d < 1.33$	score: - -
berm level	$d_b/H_{sD} \geq 0.6$	$d_b/H_{sD} < 0.6$	score: -
toe depth	$2.0 < h_t/H_{sD} \leq 2.5$	$h_t/H_{sD} > 2.5$	score: -
		$1.6 < h_t/H_{sD} \leq 2.0$	score: +
		$h_t/H_{sD} \leq 1.6$	score: ++
Fully reshaping berm breakwaters ***FR***			
lower slope	$\cot\alpha_d = 1.25/1.33$	$\cot\alpha_d < 1.25$	score: -
		$\cot\alpha_d > 1.33$	score: +
berm level	$d_b/H_{sD} \geq 0.6$	$d_b/H_{sD} < 0.6$	score: -
toe depth	$2.0 < h_t/H_{sD} \leq 2.5$	$h_t/H_{sD} > 2.5$	score: -
		$1.6 < h_t/H_{sD} \leq 2.0$	score: +
		$h_t/H_{sD} \leq 1.6$	score: ++

It is assumed that hardly and partly reshaping berm breakwaters have mostly a 1:1.5 lower slope, but fully reshaping berm breakwaters have generally a steeper slope of 1:1.25 or 1:1.33. The threshold for berm level is taken at $d_b/H_{sD} = 0.6$ for all three types of berm breakwaters. The toe depth should not have an influence on hardly reshaping berm breakwaters, compared to conventional rock layers, where damage or recession will be very limited. It certainly may have influence for partly and fully reshaping berm breakwaters and in total the toe depth has been divided in four categories.

The classification and scoring method given above can be applied to each data set individually. The recession graph of each individual data set is given in Appendix B. First the total figure for a class is given (comparable to Figure 3.14 to Figure 3.16) and then the individual data sets. The scoring of each data set, however, is made by use of Table 3.2, where the design values are given for the lower slope, berm level and toe depth. The scores are given in each graph in Appendix B and summarized in Table 3.4.

Table 3.4. Influence of lower slope, berm level and toe depth on scoring and comparison with actual recession.

Case	Type	Placement	$H_{sD}/\Delta D_{n50}$	Lower slope $\cot\alpha_d$	Berm level d_b/H_{sD}	Toe depth h_t/H_{sD}	Trend in graph	Measurement versus pprediction	Remark
Hardly reshaping					**score**				
Project 1	HR-IC	OP	1.7	0	0		**0**	**Correct**	Class II effect
L. Andersen Armour 1	HR-MA	D	1.8	-	-		**- -**	**Correct**	Deviation 1 D_{n50}
Partly reshaping									
MAST II [1996]	PR-IC	D	2.1	-	0	+	**0**	**Correct**	
Sveinbjörnsson [2008]	PR-IC	D	2.1	0	-/0	-	**0**	**More stable**	
Myhra [2005]	PR-IC	OP/D	2.0	0	0	++	**0**	**Less stable**	Class II effect
Lykke Andersen [2008]	PR-IC	D	2.1	0	-	0	**-**	**Correct**	
Project 4	PR-IC	D	2.2	-	0	++	**+**	**Correct**	
Keilisness	PR-IC	D	2.1	-	-	-	**- -**	**Correct**	Class II effect
Myhra [2005]	PR-MA	OP/D	2.0	0	0	++	**+**	**Correct**	
Project 1	PR-MA	D	2.3	-	0	0	**0**	**Correct**	
Lykke Andersen [2008]	PR-MA	D	2.2	0	-	0	**0**	**Correct**	
L. Andersen Armour 2	PR-MA	D	2.5	-	-	-	**- -**	**Correct**	Deviation 1.5 D_{n50}
Fully reshaping									
Project 1	FR-MA	D	2.9	+	-	0	**-**	**Correct**	
Project 2	FR-MA	D	2.9	0	-	++	**-**	**Less stable**	
Project 3	FR-MA	D	2.7	+	-	+/++	**0/++**	**Correct**	Deviation 2 D_{n50}
Project 4	FR-MA	D	2.8/ 3.1	0	0	++/0	**++/+**	**Correct/ More stable**	Deviation 2 D_{n50}
Project 5	FR-MA	D	3.0	0	0	-	**+**	**More stable**	
MAST II [1996]	FR-MA	D	2.9	-	0	+	**0**	**Correct**	
Moghim [2009]	FR-IC	D	2.8	0	0	++/-	**++/0**	**Correct/ More stable**	Deviation 2 D_{n50}

The three scores together on lower slope, berm level and toe depth should give an idea where the data set should be with respect to the average line in Equation 3.19. A negative total score would mean that the berm is less stable and the data points in general should lie above the recession curve. For a positive score, data points should be lower than the curve.

Table 3.4 also gives the actual trend in the data and the comparison of predicted and actual trend. Most data sets indeed are predicted correctly; a few data sets are a little less or more stable.

A few examples will be treated here. First of all, the data of Lykke Andersen [2006] give much more recession than the prediction curve, see Figure 3.21 and Figure 3.22 (Figures B.3 and B.13 in Appendix B). In both cases all scores are negative: a steep lower slope, which is less stable; a low berm, which dissipates less wave energy; and a fairly large toe depth, giving more recession, see Figure 3.20. The effect is that for hardly reshaping berm breakwaters the recession is in average $1D_{n50}$ more than predicted by Equation 3.19. For the partly reshaping berm breakwater, Figure 3.22, the deviation in average is about $1.5D_{n50}$.

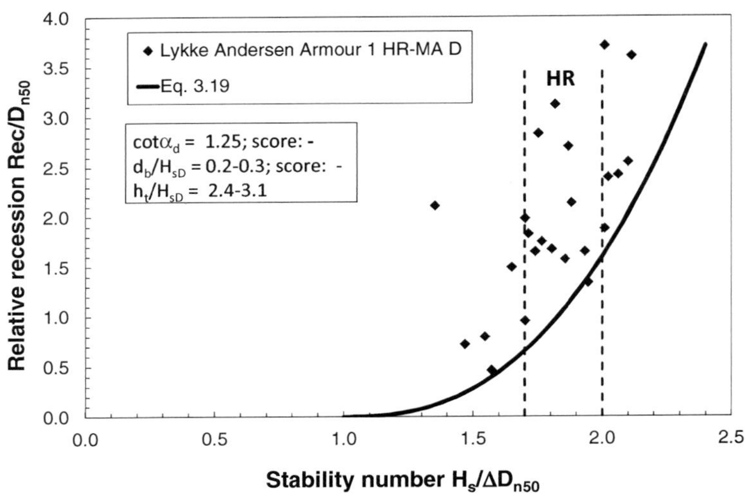

Figure 3.21. Recession data of Lykke Andersen [2006] for hardly reshaping berm breakwaters.

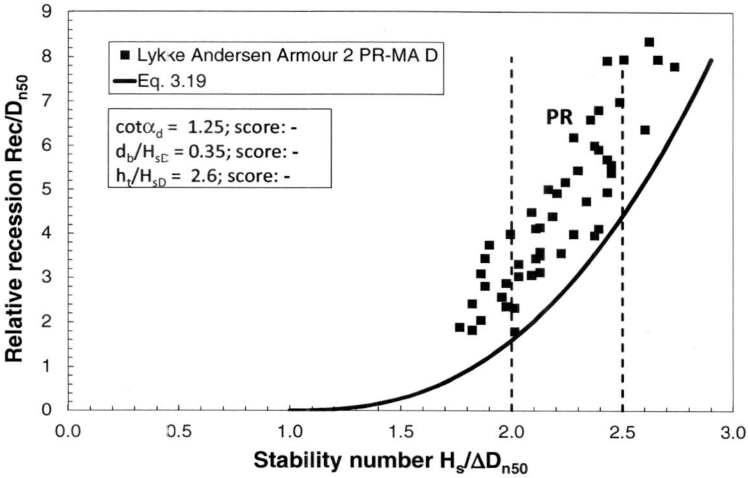

Figure 3.22. Recession data of Lykke Andersen [2006] for partly reshaping berm breakwaters.

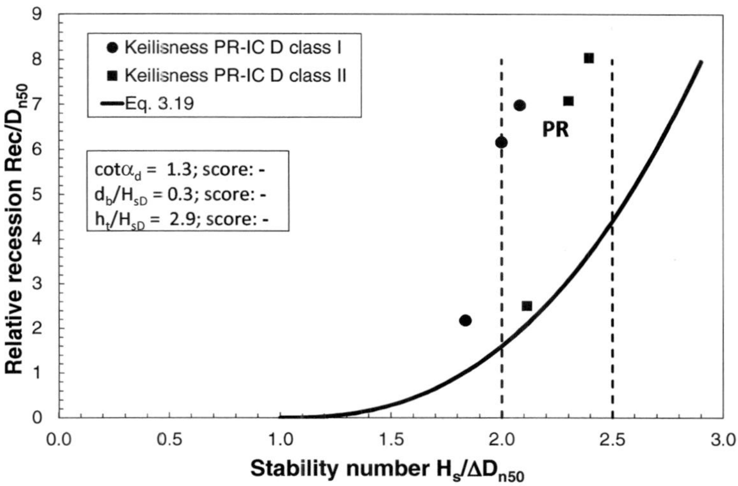

Figure 3.23. Recession data of the Keilisness project, for partly reshaping berm breakwaters.

Another example of negative scores is the Keilisness project, given in Figure 3.23. Actually, only three data points are available, but given in two ways. The structure is an Icelandic-type berm breakwater, but the large Class I rock is close to the water level and the actual behaviour of the berm must be influenced by the smaller Class II rock, which is less stable. For this reason, the dimensionless recession has been calculated for the D_{n50}'s of the Class I rock as well as the Class II rock. Even by using the smaller Class II rock the recession in the design area is more than $3D_{n50}$ larger than predicted.

The fully reshaping structure in Project 1 had a very low berm at the water level. Results are given in Figure 3.24. The overall score is neutral as the 1:1.5 slope is more stable than a 1:1.25 slope. But due to the low berm, the recession is a slightly larger than the prediction curve.

The influence of toe depth on recession is nicely demonstrated by Project 3 as well as Moghim [2009], both for fully reshaping berm breakwaters. The data of Project 3 are shown in Figure 3.25. The overall score is positive, but two data sets are shown, one with a very high toe with $h_t/H_{sD} = 1.4$. That data shows significantly less recession than with a slightly deeper toe with $h_t/H_{sD} = 1.8$.

Project 3 has another reason why it shows limited recession, see

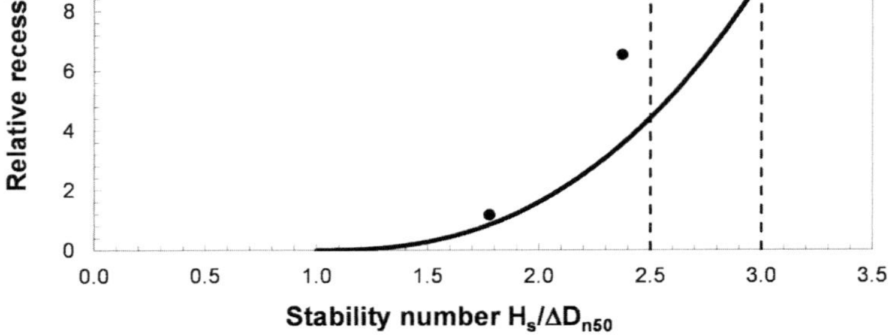

Figure 3.24. Recession data of Project 1, for fully reshaping berm breakwaters.

Figure 3.26. The berm width is quite small, but there is a considerable underwater profile almost up to half of the water depth, already creating a kind of S-shaped profile. Waves will certainly feel this profile under water and give less forces on the berm itself. Although more material is used in comparison with a smaller apron or toe, there is definitely a positive effect on stability and reshaping of the berm. The excess of material is relatively small material and less expensive than the large rock in the berm.

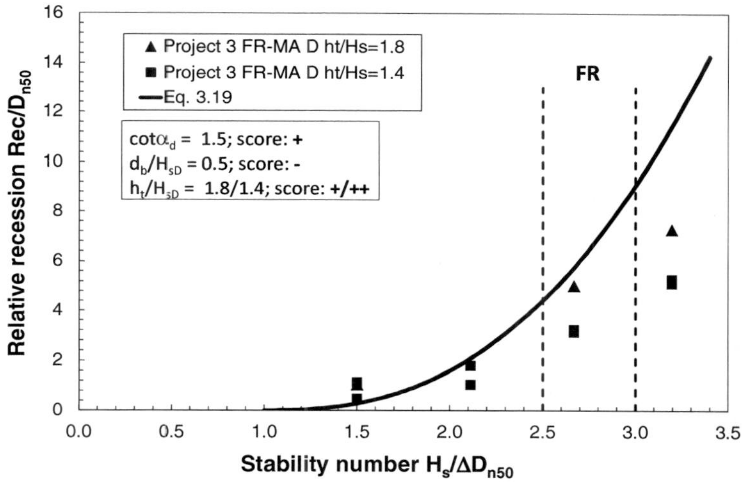

Figure 3.25. Recession data of Project 3, for fully reshaping berm breakwaters. Influence of toe depth.

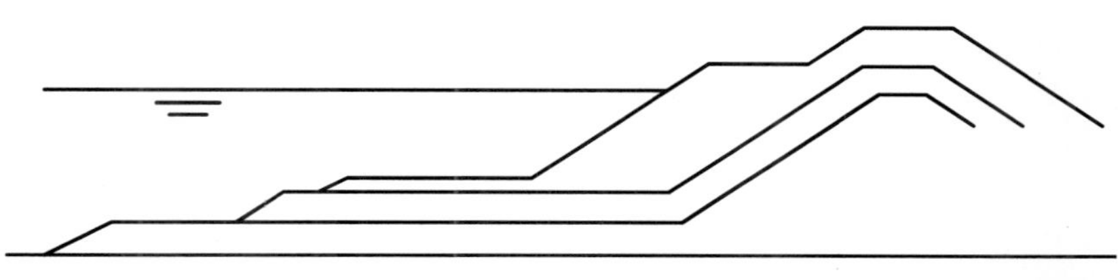

Figure 3.26. Schematic profile of berm breakwater in Project 3 with a large underwater profile, leading to less reshaping of the berm.

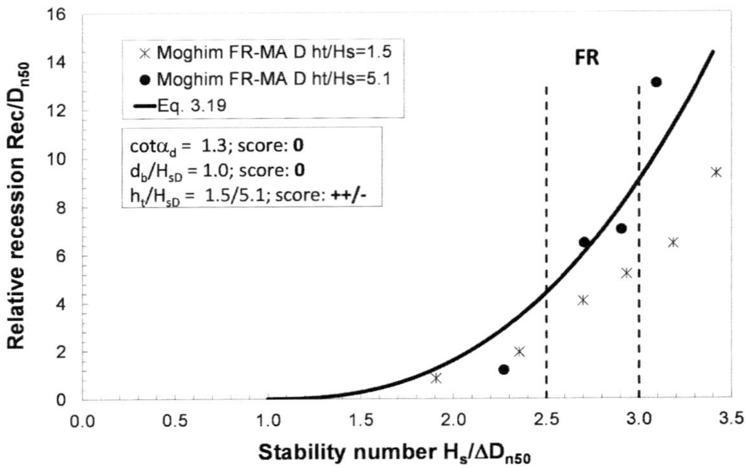

Figure 3.27. Recession data of Moghim [2009] for fully reshaping berm breakwaters.

The data of Moghim [2009] are given in Figure 3.27. The data with a small toe depth of $h_t/H_{sD} = 1.5$ is significantly more stable than the data with a deep toe. Table 3.4 shows for three cases of fully reshaping berm breakwaters that if the overall score is really positive, the recession may in average be $2D_{n50}$ less than predicted.

The scoring method applied in Table 3.4 confirms that lower slope, berm level and toe depth may have an influence on berm recession. For hardly and also partly reshaping berm breakwaters the lower slope is quite important as slopes steeper than 1:1.5 are less stable. For partly and fully reshaping berm breakwaters the toe depth has large influence on recession, certainly when the toe depth is really small.

3.7.4 *Conclusions on recession of berm breakwaters*

The main parameter to describe recession of berm breakwaters is the stability number $H_{sD}/\Delta D_{n50}$ for design conditions (100-years return period). The recession of the berm up to the design conditions can be described by Equation 3.19. The overload condition, which are conditions beyond the 100-years design condition, give a smaller increase in recession than given by Equation 3.19. Only half of the increase calculated by Equation 3.19 should be considered (see Equation 3.20).

Table 3.5. Positive and negative scores on berm reshaping, including final influence on reshaping.

Hardly reshaping berm breakwaters ***HR***			
lower slope	$\cot\alpha_d = 1.5$	$\cot\alpha_d < 1.5$	score: -
berm level	$d_b/H_{sD} \geq 0.6$	$d_b/H_{sD} < 0.6$	score: -
toe depth	h_t/H_{sD}	no influence (hardly any reshaping)	
A double negative overall score gives about $1D_{n50}$ more reshaping.			
Partly reshaping berm breakwaters ***PR***			
lower slope	$\cot\alpha_d = 1.5$	$1.33 < \cot\alpha_d < 1.5$	score: -
		$\cot\alpha_d < 1.33$	score: - -
berm level	$d_b/H_{sD} \geq 0.6$	$d_b/H_{sD} < 0.6$	score: -
toe depth	$2.0 < h_t/H_{sD} \leq 2.5$	$h_t/H_{sD} > 2.5$	score: -
		$1.6 < h_t/H_{sD} \leq 2.0$	score: +
		$h_t/H_{sD} \leq 1.6$	score: ++
A double or triple negative overall score gives about $1.5D_{n50}$ more reshaping.			
A double positive overall score gives about $1D_{n50}$ less reshaping.			
Fully reshaping berm breakwaters ***FR***			
lower slope	$\cot\alpha_d = 1.25/1.33$	$\cot\alpha_d < 1.25$	score: -
		$\cot\alpha_d > 1.33$	score: +
berm level	$d_b/H_{sD} \geq 0.6$	$d_b/H_{sD} < 0.6$	score: -
toe depth	$2.0 < h_t/H_{sD} \leq 2.5$	$h_t/H_{sD} > 2.5$	score: -
		$1.6 < h_t/H_{sD} \leq 2.0$	score: +
		$h_t/H_{sD} \leq 1.6$	score: ++
A double positive overall score, gives about $2D_{n50}$ less reshaping.			

This gives the basic information on expected recession and applies to the most common geometries of berm breakwaters. If lower slope, berm level and toe depth are different from these most common geometries, the recession could be a little more or less. The effect can be determined by a scoring method on all three influences (see Table 3.5). In that table the common geometry is given with the scoring method. The outcome of the average scoring may give the suggested deviations from the calculations with Equation 3.19 or 3.20, for design and overload conditions. Deviations will be less for conditions less than design conditions. Berm breakwaters should not be designed with stability numbers, $H_{sD}/\Delta D_{n50}$(design) > 3.0.

Chapter 4

Functional Behaviour: Wave Overtopping, Reflection and Transmission

4.1 Wave overtopping

4.1.1 *Context of wave overtopping*

Most of the guidance on wave overtopping has been given in the EurOtop Manual [2016] and one is referred to that manual for an overall picture of what wave overtopping is, what allowable overtopping could mean and prediction of mean overtopping discharges for all kind of structures. This section will mainly focus on mean wave overtopping discharges over berm breakwaters in the line of the EurOtop Manual.

The EurOtop Manual [2016] not only gives prediction formulae, but also calculation tools like PC-Overtopping and the Neural Network prediction tool. PC-Overtopping, however, is not very suitable for berm breakwaters. PC-Overtopping was developed for dike-type structures and cannot cope with a porous crest like with rock. The Neural Network prediction tool connected to EurOtop [2016], includes all berm breakwaters of Lykke Andersen [2006], using the *initial profile*. Then the nominal diameter D_{n50} with the wave height gives a kind of stability number. The stability number gives a good idea of the reshaping of the initial profile, and therefore this effect has to some extent been included in the new Neural Network. The Neural Network connected to EurOtop [2016] may be able to predict wave overtopping over berm breakwaters in a reasonable way.

Allowable overtopping should govern the design of the crest height of the berm breakwater. The EurOtop Manual [2016] gives guidance on allowable overtopping when persons are still present or vehicles are moving behind the structure. But normally for severe storms, say storms exceeding the 10-years event, nobody will be allowed to be on a breakwater. Then, the allowable overtopping depends on whether there is anything else to protect for overtopping waves or not. This might be moored ships, or access to moored ships, or infrastructure like pipelines. With respect to infrastructure it might be a good idea to make some protection against overtopping water, instead of increasing the crest height of the breakwater to such a level that overtopping does not occur. Figure 4.1 gives an example where pipelines are sheltered for forces due to wave overtopping by placing them against a crest wall.

Figure 4.1. Pipelines sheltered for wave overtopping. Akranes breakwater, Iceland. Photo by Einar Guðmundsson.

In case nothing has to be protected on the breakwater, the crest height can be optimized to the level where overtopping does not damage the crest and/or rear side too much, or where transmitted waves by overtopping will be limited to an allowable value. In these cases stability of the crest and rear side or wave transmission govern the design of the crest height. Stability of the crest and rear side has been described in Chapter 5, wave transmission in Section 4.3.

It is quite difficult to realise what allowable overtopping discharges of only 0.1 or 1 l/s per m mean. It is given as a mean discharge, where in reality the overtopping water comes by large waves reaching and overtopping the crest of the structure. But how many overtopping waves will give mean overtopping discharges of 0.1, 1 or 10 l/s per m? There may be two ways to approach this question.

Overtopping simulation in reality has been performed by the Wave Overtopping Simulator on dikes in the Netherlands, Belgium, USA and Vietnam, see Van der Meer *et al.* [2010] and [2011], Le *et al.* [2014] and Van der Meer [2014]. The simulation has been performed for assumed sea states with significant wave heights between 1–3 m. The simulations have shown that mean overtopping discharges less than 1 l/s per m do not damage dikes with grass on clay slopes and are so small that people can withstand it when standing on a dike crest. Tests with 0.1 l/s per m are often not even considered as the overtopping volumes are small and number of events too. Substantial overtopping is considered when discharges exceed 10 l/s per m.

The EurOtop [2016] website (www.overtopping-manual.com) gives access to videos of wave overtopping discharges simulated by the Wave Overtopping Simulator. Each video has a duration of 3 minutes to show how a certain overtopping discharge for a certain wave height looks. Overtopping discharges of 1, 5, 10, 30, 50 and 75 l/s per m can be chosen, for wave heights of H_s = 1 m, 2 m and 3 m, respectively. By watching some of the videos, one may get a fair idea what a certain overtopping discharge means.

The situation for berm breakwaters is a little different from dikes. Very small wave overtopping may come as splash from waves that hit individual rocks. But substantial overtopping comes as a horizontal flow over the crest, like with dikes. The main difference is the wave height considered for design conditions. Dikes do not often experience wave heights larger than 3 m, whereas berm breakwaters will often have design wave heights around 5 m. This changes the overtopping behaviour for small overtopping discharges.

Assume a design wave height for a berm breakwater around H_s = 5 m with a wave steepness of s_{op} = 0.04 (giving T_p = 8.9 s) and a storm with a duration of 6 hours. For use of PC-Overtopping or the prediction formulae in the EurOtop Manual [2016] a slope of 1:2 has been used (more or less an average slope for a berm breakwater) and an influence factor for roughness of $\gamma_f = 0.5$. Calculations show the following:

0.1 l/s per m:	no overtopping waves (splash only)
0.5 l/s per m:	5 waves overtopping, this is less than one wave per hour
1 l/s per m:	11 waves overtopping, about 2 waves per hour
5 l/s per m:	69 overtopping waves, which is 2.4% of the total incident number of waves
10 l/s per m:	139 waves, which is 4.8% of the total incident number of waves

Moreover, the prototype measurements of wave overtopping in CLASH [2004] have shown that scale effects on overtopping for rubble mound breakwaters are present if the mean discharge is less than 1 l/s per m. The discussion above and the scale effects show that it is not realistic to have allowable overtopping limits for berm breakwaters under design conditions lower than about 0.5 l/s per m. Lower discharges can be measured in a laboratory, but they have little physical meaning in reality for the circumstances described above, regardless of the frequency it may occur – design conditions or more frequent.

The EurOtop Manual [2016] gives overtopping formulae for smooth slopes like dikes, and for rubble mound structures with a straight slope. Both are considered here first as it will give the basis for describing wave overtopping for berm breakwaters.

For a sloping structure wave overtopping can be described by:

$$\frac{q}{\sqrt{g \cdot H_{m0}^3}} = \frac{0.023}{\sqrt{\tan\alpha}} \gamma_b \cdot \xi_{m-1,0} \cdot \exp\left[-\left(2.7 \frac{R_c}{\xi_{m-1,0} \cdot H_{m0} \cdot \gamma_b \cdot \gamma_f \cdot \gamma_\beta \cdot \gamma_v}\right)^{1.3}\right] \qquad 4.1$$

with a maximum of:

$$\frac{q}{\sqrt{g \cdot H_{m0}^3}} = 0.09 \cdot \exp\left[-\left(1.5 \frac{R_c}{H_{m0} \cdot \gamma_f \cdot \gamma_\beta}\right)^{1.3}\right] \qquad 4.2$$

where:

q	=	mean overtopping discharge per metre structure width	$[m^3/s$ per m]
g	=	acceleration due to gravity	$[m/s^2]$
H_{m0}	=	estimate of significant wave height from spectral analysis = $4\sqrt{m_0}$	[m]
α	=	angle between overall structure slope and horizontal	[°]
$\xi_{m-1,0}$	=	breaker parameter = $\tan\alpha/(s_{m-1,0})^{0.5}$	[-]
$s_{m-1,0}$	=	wave steepness with $L_{-m-1,0}$, based on $T_{m-1,0}$: $H_{m0}/L_{m-1,0} = 2\pi H_{mo}/(gT^2_{m-1,0})$	[-]
$T_{m-1,0}$	=	average wave period defined by m_{-1}/m_0	[s]
m_n	=	$\int_{f_1}^{f_2} f^n S(f)df$ = n^{th} moment of spectral density	$[m^{2+n}s]$
$m_{n,x}$	=	n^{th} moment of x spectral density x may be: i for incident spectrum r for reflected spectrum	$[m^{2+n}s]$
R_c	=	crest freeboard of structure	[m]
γ_b	=	influence factor for a berm	[-]
γ_f	=	influence factor for the permeability and roughness of or on the slope	[-]
γ_β	=	influence factor for oblique wave attack	[-]
γ_v	=	influence factor for a vertical wall on the slope	[-]

The wave height used for wave overtopping analysis is based on spectral analysis, H_{m0}, and not on the waves measured in the time domain ($H_{1/3}$). This is in contrast to stability of rock slopes and recession of berms in Chapter 3. It is the explicit choice in the EurOtop Manual [2016] to use this measure of wave height. The main reason was that this wave height is often the outcome of determining the design wave conditions by numerical modelling.

In many cases $H_{1/3}$ is quite similar to H_{m0}, but certainly at shallow conditions with fairly steep foreshores and small steepness, $H_{1/3}$ may become substantially larger (10%-15% is possible) than H_{m0}. One should also note that the calculation of $H_{1/3}$, and other related parameters like $H_{2\%}$, are not well described by the method of Battjes and Groenendijk [2000] if the foreshores are relatively steep (say steeper than 1:50) and the wave steepness quite low (say s_{op} smaller than 2%).

Recent tests on overtopping over a breakwater with concrete single units, [Salauddin *et al.*, 2015] showed deviations of 15% between $H_{1/3}$ and H_{m0}. This was the case with a 1:30 foreshore and a wave steepness of about $s_{op} = 0.015$. Plotting the overtopping results with H_{m0} showed large under-estimation by the overtopping formulae, where $H_{1/3}$ showed much better comparison. This is not enough proof to replace H_{m0} by $H_{1/3}$ in wave overtopping prediction, but one should be careful in using H_{m0} in conditions with steep foreshores combined with (very) low wave steepness.

The influence factors are described in detail in the EurOtop Manual [2016]. Dikes have often gentle slopes where Equation 4.1 applies. If slopes become steeper, or the wave steepness lower, Equation 4.2 may apply, which is the formula for so-called "non-breaking" or surging waves on the structure. As rubble mound breakwaters as well as berm breakwaters have steep slopes, only Equation 4.2 will be considered here, where for berm breakwaters only the lower slope (below the berm) is considered.

Actually, the dimensionless wave overtopping discharge $q/(gH_{m0}^3)^{0.5}$ in Equation 4.2 is given as a Weibull function of the relative crest freeboard R_c/H_{m0} and two influence factors, one for oblique wave attack and one for the influence of permeability or roughness of the slope. The Weibull function shows a slight curve on a log-linear graph and Figure 4.2 shows some data points for dike-type structures, where the slope was impermeable. The correct influence factors were applied on the horizontal axis, which shifts the data points to the line for a smooth straight slope under non-breaking waves.

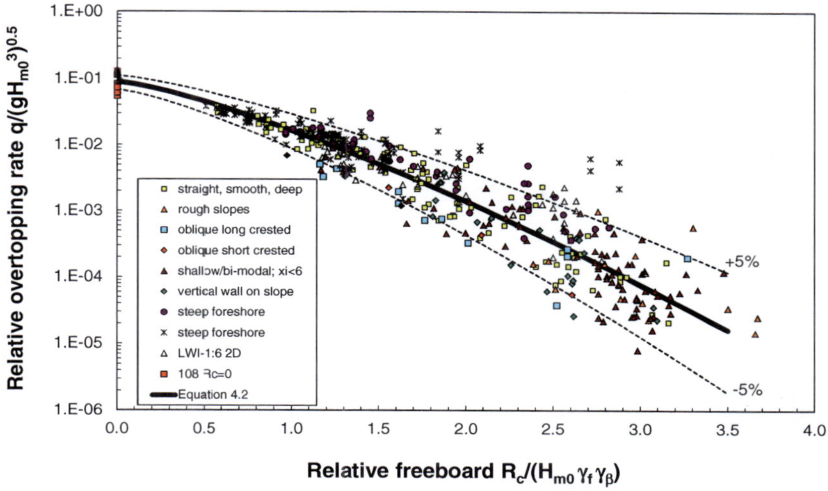

Figure 4.2. Wave overtopping for dike-type slopes under non-breaking waves, Equation 4.2. From EurOtop [2016], Fig. 5.11.

The reliability of Equation 4.1 is given by $\sigma(0.023) = 0.003$ and $\sigma(2.7) = 0.20$, and of Equation 4.2 by $\sigma(0.09) = 0.0135$ and $\sigma(1.5) = 0.15$. For a design or assessment approach it is strongly recommended to increase the average discharge by about one standard deviation. For predictions of measurements or comparison with measurements Equation 4.2 should be taken with, for instance, 5% upper and lower exceedance curves, as shown in Figure 4.2, giving the 90%-confidence interval.

Another way of showing the influence of roughness and permeability on wave overtopping is by using the relative freeboard R_c/H_{m0} on the horizontal axis and plotting different lines for different influence factors γ_f. An influence factor in fact lowers the curve for wave overtopping as shown in Figure 4.3. This figure shows the curve for $\gamma_f = 1$ for a smooth slope, similar to the curve in Figure 4.2, but it also shows curves for influence factors of $\gamma_f = 0.3$, 0.4, 0.5 and 0.6.

The data in Figure 4.3 are from CLASH [2004], see Bruce *et al.* [2009], where different armour units, including rock, were used on a slope of 1:1.5 and where overtopping was measured. Also a smooth slope was tested, giving the data points on the highest curve. Roughness and/or permeability as for rubble mound breakwaters reduce wave overtopping drastically if compared with a smooth slope. Most of the data points lay between the curves with influence factors γ_f between 0.3 and 0.6. In the EurOtop Manual [2016] a good estimate of this influence factor was made for each type of armour unit, reducing the total scatter in the figure significantly.

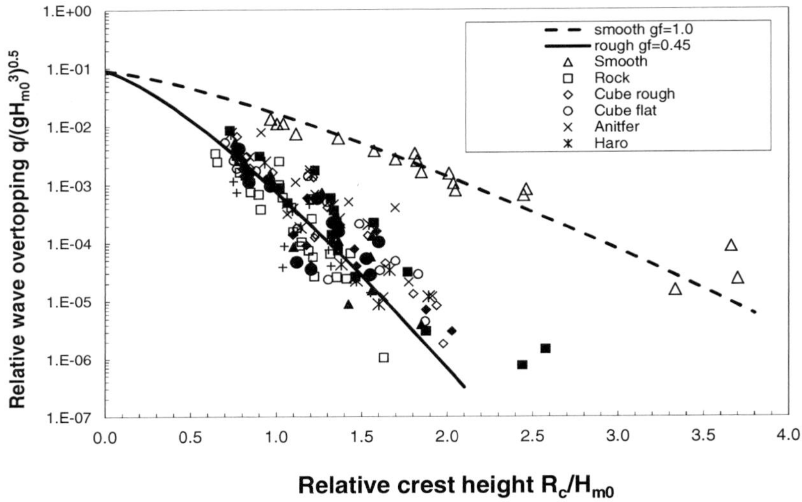

Figure 4.3. Wave overtopping for armour units on a slope of 1:1.5, Equation 4.2. From EurOtop [2016], Fig. 6.6.

A berm breakwater is also a rubble mound breakwater with large roughness and permeability. But it has not a straight slope, like the data points in Figure 4.3, but a steep seaward slope with a berm and often a partly or fully reshaped berm. Nevertheless, it may be expected that overtopping data for berm breakwaters will give a similar graph as Figure 4.3. The influence factor may then be a function of geometry of the berm breakwater or wave conditions. If this influence factor can be found and described in a sophisticated way, Equation 4.2 can be used and will place berm breakwaters in a similar graph as all other coastal structures. Equation 4.2 should be written in a slightly different way:

$$\frac{q}{\sqrt{g \cdot H_{m0}^3}} = 0.09 \cdot \exp\left[-\left(1.5\frac{R_c}{H_{m0} \cdot \gamma_{BB} \cdot \gamma_\beta}\right)^{1.3}\right] \qquad 4.3$$

where γ_{BB} is the influence factor for a berm breakwater.

In the previous EurOtop [2007] the formula of Lykke Andersen [2006] has been described, which was developed on a large systematic series of tests on berm breakwater profiles. Like his recession formula (see Chapter 3) this formula is also quite complicated, as it was based on multi-parameter fitting. In total he uses 17 parameters with 10 coefficients and 10 exponents. By such a complicated formula, it is hardly possible to find the most significant physics based influences and to make a graph like Figure 4.2 with the correct influence factor. In Chapter 2 it was concluded that there are three types of berm breakwaters: hardly reshaping, partly reshaping and fully reshaping. It may be that these types of berm breakwaters have different effects on wave overtopping. The formula of Lykke Andersen, has been described in his PhD-thesis, Lykke Andersen [2006], in Lykke Andersen and Burcharth [2010], as well as in the previous EurOtop Manual [2007]. For the reason of completeness, it has also been given here, and one may use that formula as an alternative for the formula developed in the next sections.

In order to overcome the problem that one has to calculate the reshaped profile before any overtopping calculation can be done, the formula is based on the "as built" profile, before reshaping. Instead of calculating the profile, a part of the formula predicts the influence of waves on recession of the berm. The parameter used is called f_{H0}, which is an indicative measure of the reshaping and can be defined as a "factor accounting for the influence of stability numbers". This factor has already been described for the prediction of recession of the berm in Equation 3.13. Note that f_{H0} is a dimensionless factor and not the direct measure of recession and that H_0 and T_0 are also dimensionless parameters, described by Equations 3.12 and 3.10, respectively. The parameter T_0^* has been described by Equation 3.11.

$f_{H0} = 19.8\exp(-7.08/H_0)\cdot s_{0,1}^{-0.5}$ for $T_0 \geq T_0^*$

$f_{H0} = 0.05H_0T_0 + 10.5$ for $T_0 < T_0^*$ 4.4

The berm level h_b is also taken into account as an influence factor, h_b^*. Note that the berm depth is positive if the berm level is below SWL, and therefore, for berm breakwaters often negative. This influence factor is described by:

$h_b^* = (3H_{m0} - h_b)/(3H_{m0} + R_c)$ for $h_b < 3H_{m0}$

$h_b^* = 0$ for $h_b \geq 3H_{m0}$ 4.5

The final overtopping formula then takes into account the influence factor on recession, f_{H0}, the influence factor of the berm level, h_b^*, the geometrical parameters R_c, B and G_c, and the wave conditions H_{m0} and mean period $T_{0,1}$. It means that the wave overtopping is described by a spectral mean period, not by the spectral period $T_{m-1,0}$.

$$\frac{q}{\sqrt{g \cdot H_{m0}^3}} = 1.79\,10^{-5}\,(f_{H0}^{1.34} + 9.22)\cdot s_{op}^{-2.52}\cdot \qquad 4.6$$

$$\exp\left[-5.63\left(\frac{R_c}{H_{m0}}\right)^{0.92} - 0.61\left(\frac{G_c}{H_{m0}}\right)^{1.39} - 0.55h_{b*}^{1.48}\cdot\left(\frac{B}{H_{m0}}\right)^{1.39}\right.$$

Equation 4.6 is only valid for a lower slope of 1:1.25 and an upper slope of 1:1.25. For other slopes one has to reshape the slope to a slope of 1:1.25, keeping the volume of material the same and adjusting the berm width B and for the upper slope also the crest width G_c. Note also that in Equation 4.6 the peak wave period T_p has to be used to calculate s_{op}, where the mean period $T_{0,1}$ has to be used in Equation 4.4.

Although no tests were performed on the non-reshaping Icelandic berm breakwaters, a number of tests were performed on non-reshaping structures by keeping the material in place with a steel net. The difference may be that Icelandic berm breakwaters show a little less overtopping, due to the presence of larger rock and, therefore, more permeability. The tests showed that Equation 4.6 is also valid for non-reshaping berm breakwaters, if the reshaping factor $f_{H0} = 0$. For more background on the formulae, see Lykke Andersen, [2006].

4.1.2 *Available data sets*

Data sets on berm recession have been described in Chapter 3 with an overall view in Table 3.2. Wave overtopping was only measured in a number of these data sets and these are given in Table 4.1

Table 4.1. Overall view of data used for overtopping analysis.

Case	Type	$H_{sD}/\Delta D_{n50}$	Lower slope $\cot\alpha_d$	Upper slope $\cot\alpha_u$	Berm width B/H_{sD}	Berm level d_b/H_{sD}	Crest height R_c/H_{sD}	Number of tests $q/(gH_s^3)^{0.5}>10^{-5}$
Hardly reshaping								
Project 1	HR-IC	1.7	1.5	1.5	1.5	0.9	1.6	8
Lykke Andersen Armour 1	HR-MA	1.8	1.25	1.25	0-4	0.2-0.3	0.8-1.2	32
Partly reshaping								
Project 4	PR-IC	2.2	1.25	3.0	3.0	0.9	1.3	7
Lykke Andersen [2008]	PR-IC	2.0	1.5/1.3	1.5	1.5	0.6	1.3/1.7	10
Project 1	PR-MA	2.3	1.1	1.5	2.6	0.9	1.4	2
Project 1	PR-MA	2.3	1.1	1.5	4.3	0.9	1.4	2
Project 1	PR-MA	2.3	1.1	1.5	2.7	1.6	1.4	1
Lykke Andersen Armour 2	PR-MA	2.5	1.25	1.25	2.5-4	0.35	0.8-1.5	27
Lykke Andersen [2008]	PR-MA	2.2	1.5/1.3	1.5	1.5	0.6	1.3	5
Keilisnes	PR-MA	2.1	1.3	2.25	2.9	0.3-0.4	1.65	15
Full reshaping								
Project 4	FR-MA	2.8	1.25	3.0	2.5	0.9	1.5	7
Project 4	FR-MA	3.1	1.25	1.5	3.0	0.7	1.5	6
Project 5	FR-MA	3.5	1.33	1.33	2.3	0.6/1.0	0.6/1.0	4
Project 5	FR-MA	3.0	1.33	1.33	3.1	0.6/1.0	0.6/1.0	8
Lykke Andersen Armour 3	FR-MA	3.5	1.25	1.25	2.5-5	0.35	0.8-1.5	54
Dynamically stable (not considered)								
Lissev [1993]	FR-MA	4.5	1.25	1.5	3.0	0.4	1.3	0

Abreviations in the table

HR-IC	Hardly reshaping Icelandic-type berm breakwater
HR-MA	Hardly reshaping mass-armoured berm breakwater
PR-IC	Partly reshaping Icelandic-type berm breakwater
PR-MA	Partly reshaping mass-armoured berm breakwater
FR-MA	Full reshaping mass-armoured berm breakwater
H_{sD}	Estimated 100-years design wave height or second highest wave height in a research series

The partly reshaping mass-armoured breakwater in Project 1 had three specific tests: one with a much longer berm and another with a high berm at the same level as the crest. They are given separately in Table 4.1. Project 5 had the berm level at the same level as the crest, but it was quite a low crest level, more according to a normal berm level. In Lykke Andersen *et al.* [2008] one cross-section was tested, but rock gradings were combined to very wide gradings giving an Icelandic-type berm breakwater, a hybrid berm breakwater (a few gradings, in between an Icelandic-type of berm breakwater and a fully reshaping berm breakwater with only two rock classes) and a fully reshaping berm breakwater. Due to its wide grading, the fully reshaping berm breakwater cross-section is completely outside a normal design range and will not be considered here. The first cross-section is considered as a partly reshaping Icelandic-type berm breakwater (PR IC) and the hybrid is considered as a partly reshaping mass-armoured berm breakwater (PR MA). Also here some tests were outside the realistic range with wave heights up to 11 m, where the design wave height was assumed to be around 7 m.

Wave overtopping has not been measured for many fully reshaping berm breakwaters. It is for this reason that also some tests on armour 3 of Lykke Andersen [2006] have been considered, where recession results were not dynamically stable (actually most profiles are dynamically stable). The "design" stability number was $H_s/\Delta D_{n50} = 3.5$. Overtopping for larger stability numbers were not considered. Also the tests of Lissev [1993] were not considered as the stability number of $H_s/\Delta D_{n50} = 4.5$ is far beyond the limit of statically stable.

Overtopping in a laboratory can be measured very accurately, but the meaning of very small overtopping is not always realistic (see also Section 4.1.1). Overtopping rates lower than 1 l/s per m are affected by scale effects. Wave overtopping graphs are given in relative form, using $q/(gH_s^3)^{0.5}$ as dimensionless overtopping rate. It depends on the significant wave height and what the actual overtopping rate will be if a certain threshold for the relative overtopping rate is taken. Assume a threshold of $q/(gH_s^3)^{0.5} = 10^{-5}$. Then the following actual overtopping rates can be calculated:

$H_s = 1$ m	$q = 0.03$ l/s per m
$H_s = 4$ m	$q = 0.25$ l/s per m
$H_s = 5$ m	$q = 0.35$ l/s per m
$H_s = 6$ m	$q = 0.46$ l/s per m

A value of $q/(gH_s^3)^{0.5} < 10^{-5}$ will therefore generally give overtopping less than 0.5 l/s per m. This is already a threshold where scale effects are likely to play a role. For analysis of overtopping data the focus will therefore be on $q/(gH_s^3)^{0.5} > 10^{-5}$. For smaller values the scatter also increases. In the overtopping graphs the area below the given threshold will be shaded in a way that small overtopping data are still visible, but the graphs make also clear that the focus is on larger overtopping rates rather than the very small ones.

Lykke Andersen [2006] performed almost 600 tests on wave overtopping, but a large part of them are outside any practical design range (see below) and in some cases overtopping conditions were generated on already reshaped berm breakwaters, but now with lower wave conditions. In order to compare the data with the other data from real projects, the data set has been reduced. For Armour 1, the hardly reshaping mass-armoured berm breakwater, HR-MA, 214 tests on overtopping were performed. On only 36 cross-sections the recession was measured and these are considered closest to design conditions. All these tests had overtopping rates with $q/(gH_s^3)^{0.5} > 10^{-5}$. Four tests with a berm below SWL were not taken into account, which brings the total number of tests to 32.

Armour 2 can be described as a partly reshaping mass-armoured berm breakwater. In total, 57 tests were performed with 27 tests with $q/(gH_s^3)^{0.5} > 10^{-5}$. Armour 3 consisted of fully reshaping mass-armoured cross-sections (FR MA). Many of them had large berms or stability numbers beyond static stability. FR means fully reshaping where no or only a small part of the original berm has left. Tests where more than one significant wave height on berm width was left were not considered. This means roughly that B-Rec < 0.1 m. In the total of 323 tests that were performed on Armour 3 structures, 115 of them had "too large berms" and 139 test had stability numbers $H_s/\Delta D_{n50} > 3.5$. From the remaining tests, 13 had overtopping rates $q/(gH_s^3)^{0.5} < 10^{-5}$, which brings the total number of tests considered for Armour 3 to 54.

The overtopping data from site specific projects are then 75 tests, where for Lykke Andersen [2006] 113 tests remain.

4.1.3 *Development of influence factor γ_{BB}*

Each data set was separately plotted in a graph with relative wave overtopping rate versus relative crest freeboard. The data was compared with Equation 4.2 for steep slopes, with various values for the influence factor γ_f = 1.0 (smooth slope); 0.6; 0.5; 0.4 and 0.3. For berm breakwaters one could also read $\gamma_f = \gamma_{BB}$.

It is known that the wave period has influence on overtopping at berm breakwaters, which is in contrast to steep smooth slopes and also to conventional breakwaters with a straight and steep slope. The reason may be the berm itself, which is very permeable and is most effective for dissipation of energy of short waves. For this reason, each test on overtopping was classified into a wave steepness range, given by s_{op} = 0.005–0.01; 0.01–0.02; 0.02–0.03; 0.03–0.04; 0.04–0.05 and 0.05–0.06. All individual graphs are given in Appendix C. Here only two are taken as an example, the data of Keilinsnes and of Lykke Andersen [2006] on Armour 3.

Figure 4.4 shows the overtopping data of Keilisnes, [Sigurdarson and Viggosson, 1994], a partly reshaping Icelandic-type berm breakwater. Although half of the data lies below the threshold, it is very clear that lower wave steepness (larger wave periods) give larger wave overtopping.

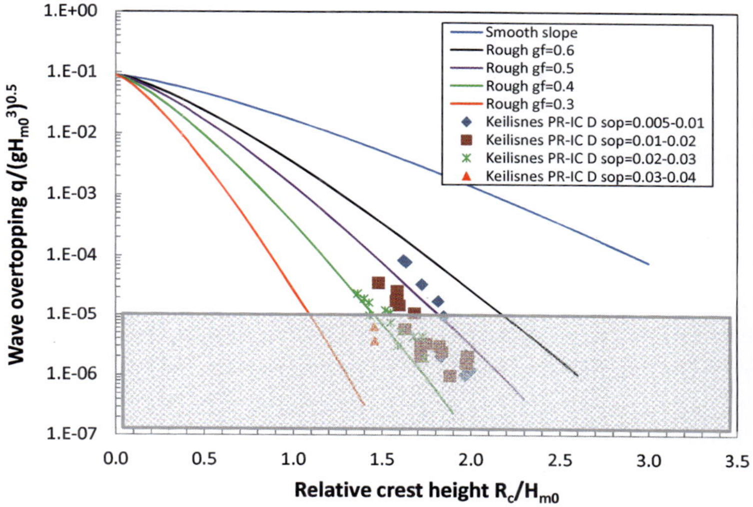

Figure 4.4. Wave overtopping for Keilisnes, PR IC.

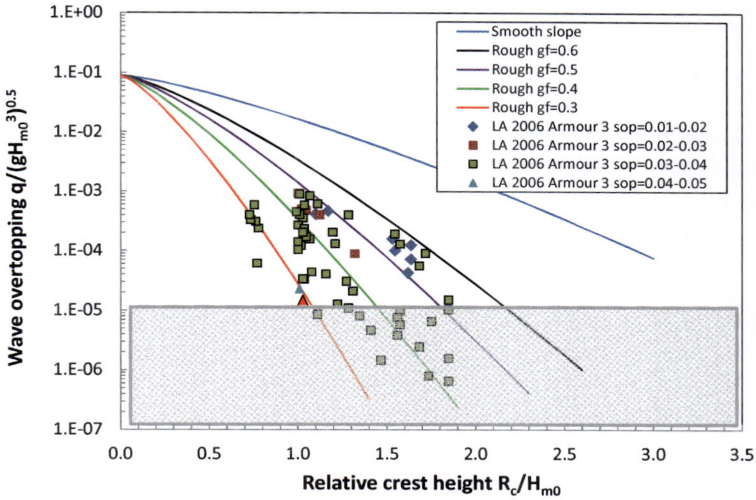

Figure 4.5. Wave overtopping for Lykke Andersen [2006], armour 3, FR MA.

In Figure 4.5, with the data of Armour 3 of Lykke Andersen [2006] there is again a clear influence of the wave steepness. A large scatter is present for steepness $s_{op} = 0.03$-0.04, where a large number of tests were performed with different berm widths, crest heights and crest widths. The structure was a fully reshaping mass-armoured berm breakwater FR MA.

Other data sets also show to greater of lesser extent the influence of wave steepness. Only the data set of Project 5 shows no influence of wave steepness at all and all data points seem to be on one line (see Appendix C). This is the data set where the crest level was similar to the berm level. An explanation for this different behaviour has not been found.

The general conclusion of analysing the various graphs is that there is a clear influence of the wave steepness, although this may be still different for the different types of berm breakwaters: HR, PR and FR.

The next step of analysis was to determine for each data set and for each class of wave steepness a value (or often a range of values) for the influence factor γ_{BB}. These ranges of γ_{BB} are given in Figure 4.6 and Figure 4.7 as a function of the wave steepness. Figure 4.6 combines the data sets for HR and PR, where Figure 4.7 gives the ranges for the fully reshaping structures, FR. In both cases, the trend is quite clear that the influence factor decreases with increasing steepness (giving decreasing overtopping). Note that the data in Figure 4.6 and Figure 4.7 show ranges and not actual data. A fitting line between the upper and lower boundary would always be quite good. The data of Project 5 show indeed no influence of the wave steepness (see Figure 4.7), and give an opposite trend than the other two data sets.

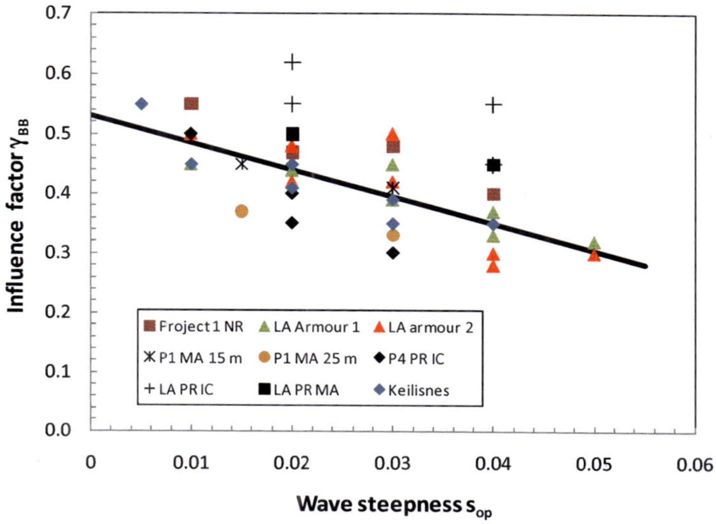

Figure 4.6. Influence factors γ_{BB} given as a range for each class of wave steepness. HR and PR.

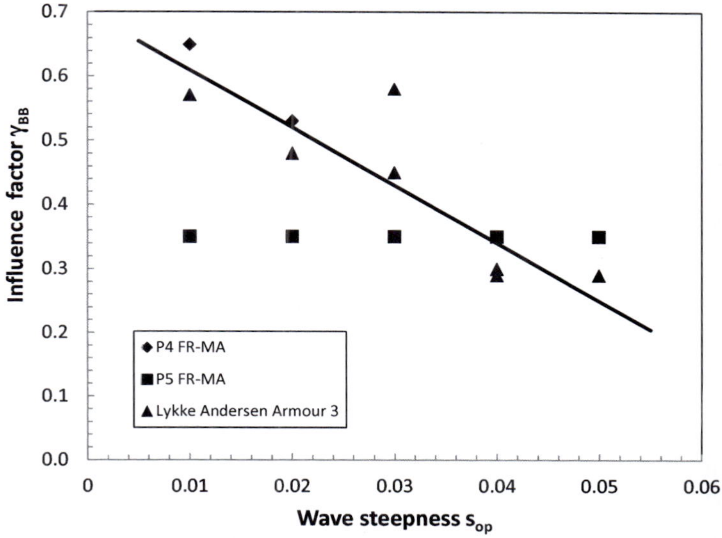

Figure 4.7. Influence factors γ_{BB} given as a range for each class of wave steepness. FR.

It seems that the influence of wave steepness on wave overtopping for berm breakwaters can be given by a linear trend:

$$\gamma_{BB} = 0.53 - 4.5\ s_{op} \qquad \text{for HR and PR} \qquad 4.7$$

$$\gamma_{BB} = 0.70 - 9.0\ s_{op} \qquad \text{for FR} \qquad 4.8$$

Equation 4.8 is quite good for fully reshaping berm breakwaters in Figure 4.7, except of course for Project 5. This equation can be seen as the end result of analysis for this kind of structure. In Figure 4.6, however, a number of ranges are completely outside the given trend line. Some of them are located higher (data sets Project 1 HR, Lykke Andersen PR IC) and others are clearly lower than the line (data sets Project 1 MA 25 m wide berm, Project 4 PR IC). A closer look at the data reveals that the berm width may have an effect: the data above the trend have in general small berms with $B/H_{sD} = 1.5$, where the data below have wide berms up to $B/H_{sD} = 4.3$.

The following step was to determine for each range of γ_{BB}'s in Figure 4.6 the difference with the trend line Equation 4.7, given as $\Delta\gamma_{BB}$. This difference is then plotted against the relative berm width B/H_s, as defined in Table 4.1, which means using the 100-years wave height as the wave height to classify the structure, or the 80% value of the maximum wave height in a series of research tests. It is *not* the wave height used in the test, as Figure 4.6 gives ranges for different test conditions, where wave heights for individual sub tests have been lost. Figure 4.8 gives the differences with the trend line Equation 4.7 and also here ranges are given.

A trend line through the data ranges can be given by:

$$\Delta\gamma_{BB} = 0.15 - 0.05B/H_{sD} \qquad \text{for HR and PR} \qquad 4.9$$

Note that in Equation 4.9 $\Delta\gamma_{BB} = 0$ if $B/H_{sD} = 3$, which means that such a berm width under design conditions can be described by Equation 4.7 only. Smaller berm widths give a positive $\Delta\gamma_{BB}$ (more overtopping) and a negative $\Delta\gamma_{BB}$ gives less overtopping for larger berm widths.

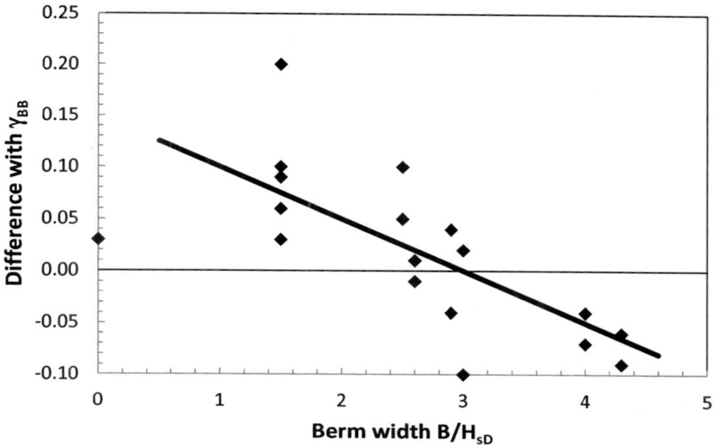

Figure 4.8. Difference with Equation 4.4, for HR and PR berm breakwaters, given as a function of relative berm width B/H_{sD}. Note that H_{sD} is the "design" wave height for the structure.

The final formula for γ_{BB} for HR and PR berm breakwaters can then be described by combination of Equations 4.7 and 4.9:

$$\gamma_{BB} = 0.68 - 4.5s_{op} - 0.05B/H_{sD} \qquad \text{for HR and PR} \qquad 4.10$$

Equations 4.8 and 4.10 give the final formulae for the influence factor on γ_{BB}, separated in two classes of berm breakwaters: the hardly and partly reshaping berm breakwaters and the fully reshaping berm breakwaters. The first class is influenced by the wave steepness as well as the relative berm width under design conditions, the second one is influenced only be wave steepness. There may be more influences (see Lykke Andersen [2006]), but these are the most significant ones.

The influence factors given by Equations 4.8 and 4.10 can be introduced in the relative crest height as $R_c/(H_s\ \gamma_{BB})$ and again overtopping graphs can be made where scatter now should be reduced. Appendix C gives all the graphs for individual datasets. Note that by introducing γ_{BB} in the relative crest height on the horizontal axis, all data points for berm breakwaters shift to the right and can now directly be compared with the curve for a "smooth slope", as that is the basis for the overtopping equation, see Equation 4.3.

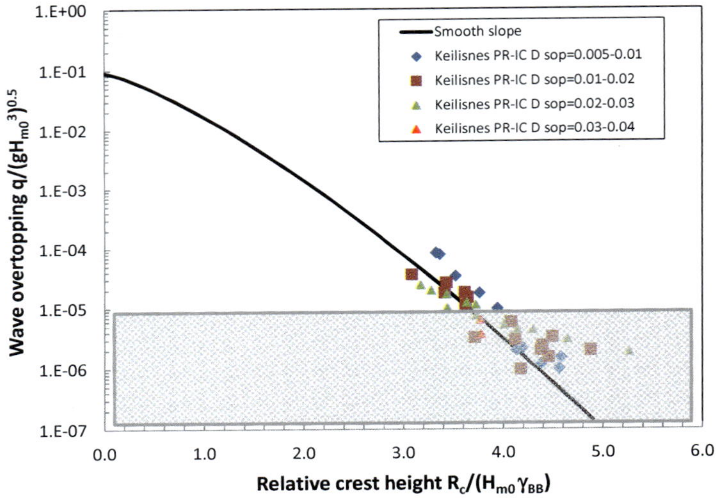

Figure 4.9. Wave overtopping for Keilisnes, including γ_{BB}, PR IC.

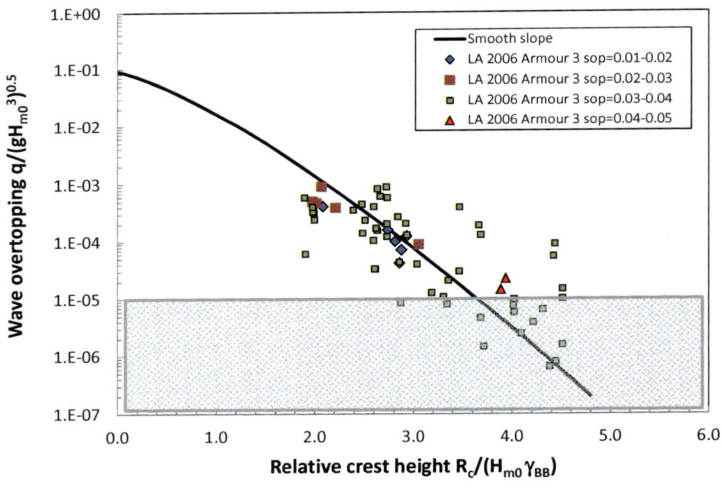

Figure 4.10. Wave overtopping for Lykke Andersen [2006] armour 3, including γ_{BB}, FR MA.

Here the same examples of Keilisnes [Sigurdarson and Viggosson, 1994] and Lykke Andersen [2006] Armour 3 are given in Figure 4.9 and Figure 4.10. In both cases the data are nicely around the line, although part of the Lykke Andersen test data still shows a significant scatter.

In general the graphs in Appendix C show much less scatter by using γ_{BB} and are more or less grouped around the prediction line. The outlier remains of course Project 5.

Figure 4.11 and Figure 4.12 give then the final result with all data combined into these two graphs. The prediction line is given together with the 5% exceedance curves. The vast majority of the data lies indeed between these two exceedance curves. Project 5 may be the main outlier with half of the data outside the confidence band.

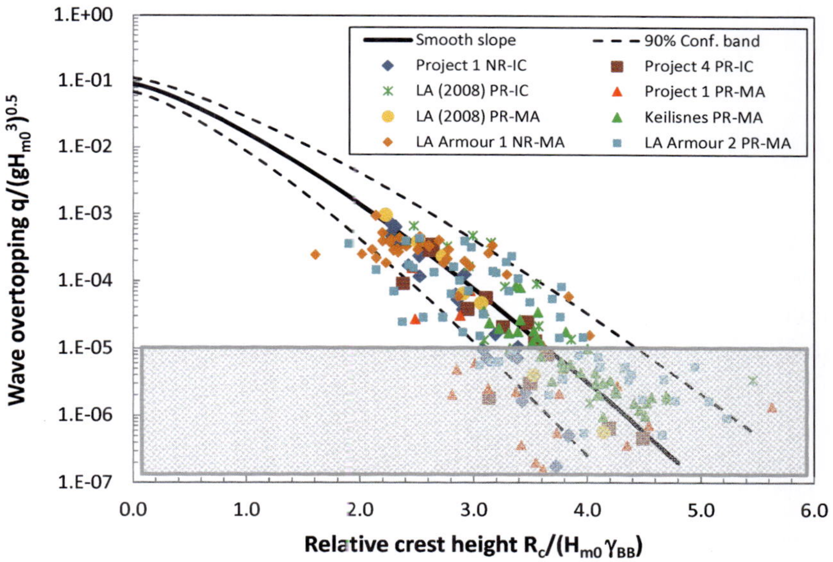

Figure 4.11. Wave overtopping for hardly and partly reshaping berm breakwaters, HR and PR.

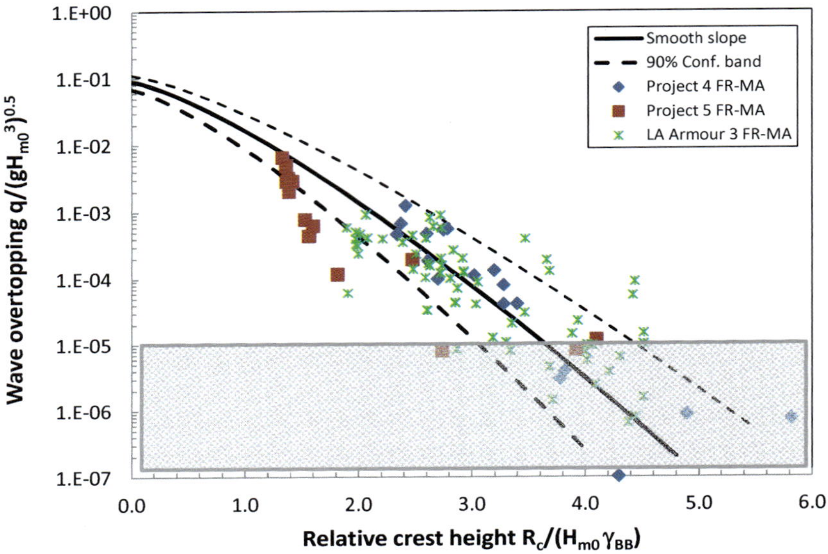

Figure 4.12. Wave overtopping for fully reshaping berm breakwaters, FR.

It means that with the influence factors derived in this section the general overtopping formula for steep smooth and rough slopes can also represent overtopping of berm breakwaters with similar reliability. Summarizing, overtopping for berm breakwaters can be calculated by:

$$\frac{q}{\sqrt{g \cdot H_{m0}^3}} = 0.09 \cdot \exp\left[-\left(1.5\frac{R_c}{H_{m0} \cdot \gamma_{BB} \cdot \gamma_\beta}\right)^{1.3}\right] \qquad 4.3$$

with:

$$\gamma_{BB} = 0.68 - 4.5 s_{op} - 0.05 B/H_{sD} \qquad \text{for HR and PR} \qquad 4.11$$

$$\gamma_{BB} = 0.70 - 9.0\ s_{op} \qquad \text{for FR} \qquad 4.12$$

and B/H_{sD} is given by the design wave height.

For $q/(gH_s^3)^{0.5} < 10^{-5}$ formulae 4.3, 4.8 and 4.10 may over-predict the actual wave overtopping. But scale effects may bring the overtopping up again. Mostly a safe design is created if the formulae are also used for $q/(gH_s^3)^{0.5} < 10^{-5}$; in any case, one should be careful in interpreting model test results for these low values.

Equations 4.3, 4.8 and 4.10 give the so-called *mean value approach*, see also EurOtop [2016]. The equations give the mean of the data points, and this method should be used to compare with measurements and it can be used in a probabilistic approach if the equations are rewritten to a reliability function, taking into account the given standard deviations. For a *design or assessment approach*, however, it is better to take a little safety into account as the equations give quite some scatter. A good way is to include one standard deviation extra into this design or assessment approach, which gives the following design formulae:

$$\frac{q}{\sqrt{g \cdot H_{m0}^3}} = 0.1035 \cdot \exp\left[-\left(1.35 \frac{R_c}{H_{m0} \cdot \gamma_{BB} \cdot \gamma_\beta}\right)^{1.3}\right] \qquad 4.13$$

with:

$$\gamma_{BB} = 0.68 - 4.5 s_{op} - 0.05 B/H_{sD} \qquad \text{for HR and PR} \qquad 4.14$$

$$\gamma_{BB} = 0.70 - 9.0\, s_{op} \qquad \text{for FR} \qquad 4.15$$

and B/H_{sD} is given by the design wave height.

Finally, a warning should be given in application of the given overtopping formulae in shallow water with relatively steep foreshore slopes. The wave height to be used is the incident significant wave height H_s at the toe of the structure. For wave overtopping the spectral wave height, $H_{m0} = 4(m_0)^{0.5}$, has to be taken, following CLASH [2004] and EurOtop [2016]. Another definition of significant wave height is the average of the highest third of the waves, $H_{1/3}$. This wave height $H_{1/3}$ is, in principle, the wave height that should be used in the Van der Meer formulae on stability of rock slopes [Van der Meer, 1988-a] and has also been used in Chapter 3 to describe berm recession. In deep water, both definitions produce almost the same value, but situations in shallow water and fairly steep foreshore slopes can lead to differences of 10–15%.

Recent overtopping tests on a single layer concrete armoured breakwater with a 1:30 foreshore slope, indeed showed these differences in wave height (Salauddin *et al.*, [2015]). It was clearly shown that the performance with overtopping equations became much better by using the (larger) $H_{1/3}$ than with the spectral wave height H_{m0}. There is not enough proof to change to $H_{1/3}$ in overtopping equations, but one should be aware that in situations with fairly steep foreshores and wave shoaling and breaking, overtopping may significantly be under-predicted by using the wave height H_{m0}.

4.2 Wave reflection

The stability of a breakwater and the crest height design are often the main aspects in the design procedure for a breakwater. For a berm breakwater it means the description of the recession of the berm (and damage if the recession is very limited), as described in Chapter 3, and the calculation of overtopping or the calculation of the crest height given allowable overtopping (Section 4.1). Another design aspect may be wave reflection. Large wave reflection may be unwanted as it may hinder ship navigation or increase erosion of adjacent beaches.

In general wave reflection for berm breakwaters is fairly low, comparable to or lower than for conventional rock structures. But measuring wave reflection is often a by-product of physical model tests to establish incident significant wave heights, so data are available to give a prediction of reflection for berm breakwaters.

Not all data sets described in Chapter 3 or Section 4.1 included the reflection coefficient K_r, where $K_r = H_{m0,i}/H_{m0,r}$ (with $H_{m0,i}$ = incident significant wave height and $H_{m0,r}$ = reflected significant wave height). Reflection data were only available from Project 1 (HR), Project 4 (PR and FR), Sveinbjörnsson [2008]—PR, Myhra [2005]—PR, MAST II [1996] —PR and FR and Lykke Andersen [2006]—HR, PR and FR, where HR, PR and FR are hardly, partly and fully reshaping berm breakwaters.

Most reflection formulae are given as function of the breaker parameter $\xi = \tan\alpha/s^{0.5}$, where $\tan\alpha$ = slope angle and s = wave steepness.

An overall view of all kind of rubble mound and smooth structures is given in Zanuttigh and Van der Meer [2008]. Also composite slopes are given in that reference, where an average slope has been defined to be used as tanα.

The problem with berm breakwaters is that first the bermed structure is in fact a composite slope, but this slope also changes more or less during recession. A fully reshaping berm breakwater finally becomes a nice S-shaped profile, which is different from a structure with straight slopes. This means that it is very difficult to follow the conventional method with the breaker parameter as in Zanuttigh and Van der Meer [2008]. An alternative method will be given here.

The conventional way of analysis uses the slope angle as well as the wave steepness in the breaker parameter. Data for other breakwaters show that the wave steepness has a significant influence on wave reflection. Given this fact and the problem in establishing an average slope for a reshaping berm breakwater, a first analysis can be done for the wave steepness only. Figure 4.13 gives an overall view of all available reflection data for statically stable berm breakwaters with design stability numbers $H_s/\Delta D_{n50} < 3.0$

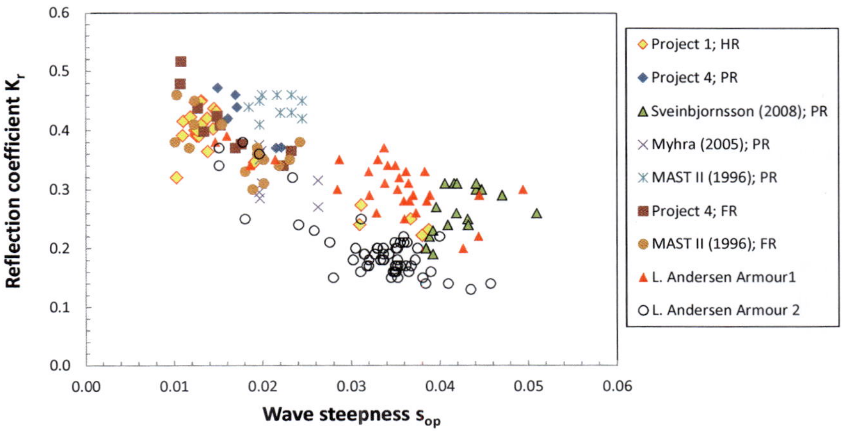

Figure 4.13. All available data on reflection for statically stable berm breakwaters as function of wave steepness.

Two conclusions can be drawn from the figure. First that wave reflection decreases with increasing wave steepness, which is according to other breakwaters, [Zanuttigh and Van der Meer, 2008]. The second conclusion is that there is quite a lot of scatter if all data are taken together in one graph. Further analysis has therefore been focused below on explaining the scatter.

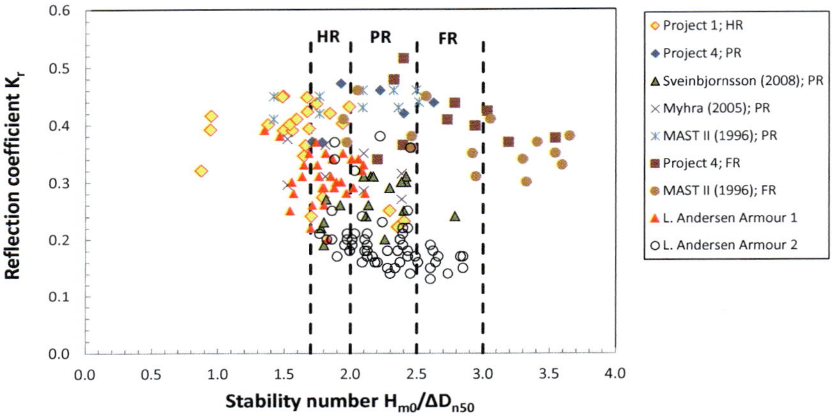

Figure 4.14. All available data on reflection for statically stable berm breakwaters as function of the stability number.

Figure 4.14 shows again all reflection data, but now as a function of the stability number $H_s/\Delta D_{n50}$. The design areas for hardly, partly and fully reshaping are also given in the graph. Most of the data are in the areas for hardly and partly reshaping, except for the fully reshaping data of Project 4 and MAST II [1996]. Also a part of Lykke Andersen [2006] data with Armour 2 are in the fully reshaping area. As described in Chapter 3, this data with Armour 2 (and also Armour 1) gave more recession than the average trend for berm breakwaters, mainly due to a steep lower slope, combined with a fairly low berm and relatively deep water. As these data gave more reshaping than most partly reshaping berm breakwaters, its performance looked more like that of a fully reshaping berm breakwater. For this reason, a similar graph was made in Figure 4.15, but now with reflection as a function of dimensionless recession Rec/D_{n50}.

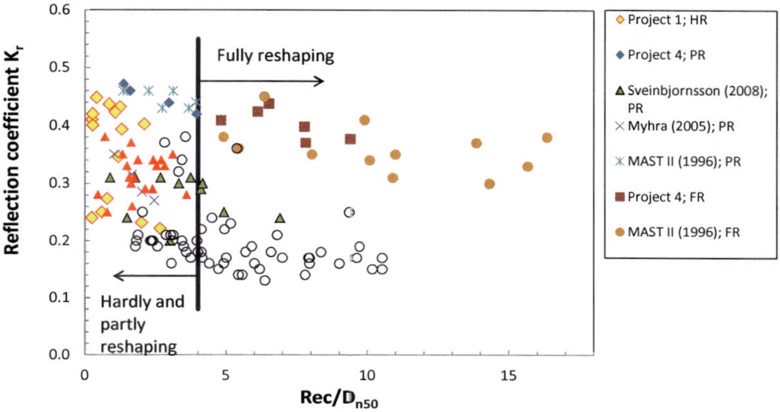

Figure 4.15. All available data on reflection for statically stable berm breakwaters as function of the dimensionless recession.

If the transition between partly and fully reshaping berm breakwaters is taken around Rec/D_{n50} = 4–5, then indeed the majority of data of Lykke Andersen [2006] with armour 2 belong to the area of fully reshaping berm breakwaters. Of course there still is a tremendous scatter on both sides of the line in Figure 4.15, but it is possible to divide the data between hardly or partly reshaping berm breakwaters and fully reshaping berm breakwaters.

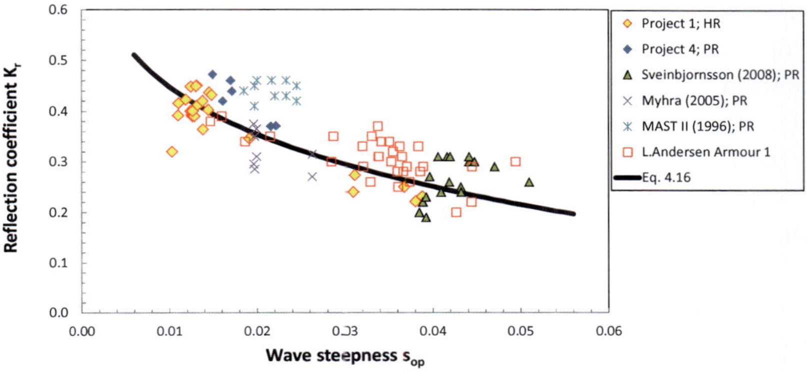

Figure 4.16. Data on reflection for hardly and partly reshaping berm breakwaters as function of wave steepness.

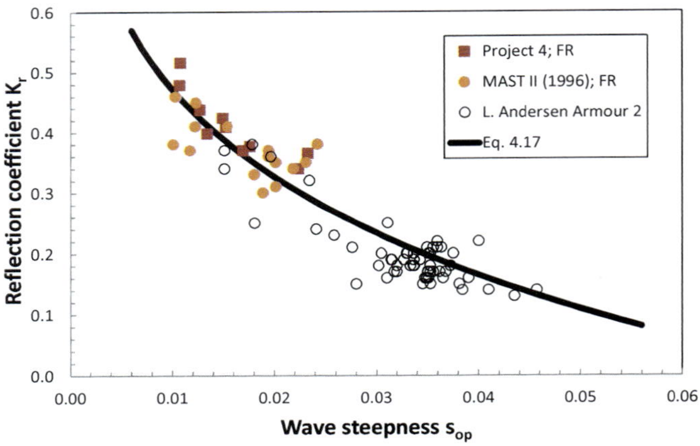

Figure 4.17. Data on reflection for fully reshaping berm breakwaters as function of wave steepness.

Two graphs have been made, Figure 4.16 and Figure 4.17, with the two categories of berm breakwaters. Both figures show much less scatter than in the overall picture in Figure 4.13 and they show a nice trend as function of the wave steepness. Part of the scatter in Figure 4.16 could be explained by the different "as built" lower slopes, where a steep slope of 1:1 or 1:1.25 (MAST II and Lykke Andersen data) give slightly more reflection than a slope of 1:1.5 (Project 1). However, as a first estimate, the trends in Figure 4.16 and Figure 4.17 are good enough for a prediction.

In Section 4.1 on wave overtopping the Armour 3 data of Lykke Andersen [2006] were considered, although the design stability number of $H_s/\Delta D_{n50}$ =3.5 is in the area of dynamic stability. It means that there has been quite a lot of reshaping to fully developed S-shaped profiles for the majority of the wave conditions. Nevertheless, it is interesting to compare these data with the data on fully reshaping berm breakwaters as in Figure 4.17. Figure 4.18 gives the comparison.

Majority of the data gives similar reflection as for fully reshaping berm breakwaters. Only for wave steepness s_{op} < 0.025 are reflections for the dynamically stable structures a little lower.

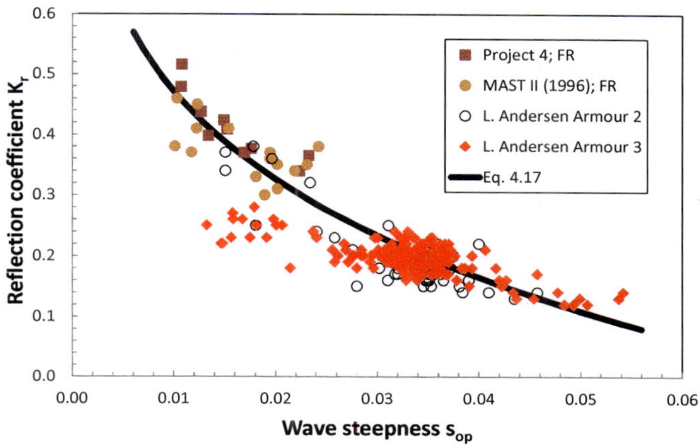

Figure 4.18. Data on reflection for fully reshaping berm breakwaters as function of the wave steepness, including Armour 3 data of Lykke Andersen [2006] for dynamically stable structures.

Wave reflection for berm breakwaters can be calculated by the following prediction formulae:

For hardly and partly reshaping berm breakwaters ($H_{sD}/\Delta D_{n50}$ <2.5 or Rec/D_{n50} < 4–5):

$$K_r = 1.3 - 1.7 s_{op}^{0.15} \qquad 4.16$$

Steeper as built downward slopes than 1:1.5 may give a little more reflection.

For fully reshaping berm breakwaters ($H_{sD}/\Delta D_{n50}$ >2.5 or Rec/D_{n50} > 4-5):

$$K_r = 1.8 - 2.65 s_{op}^{0.15} \qquad 4.17$$

4.3 Wave transmission

Waves overtopping a structure may generate waves behind that structure, which is called wave transmission. Low-crested structures may give severe wave transmission, but most breakwaters in general give only limited wave transmission during design conditions. The wave transmission coefficient K_t is defined as $K_t = H_{m0,t}/H_{m0,i}$, where $H_{m0,t}$ and $H_{m0,i}$ are transmitted and incident wave height, respectively.

Three of the data sets considered earlier in this chapter also gave measurements of wave transmission. The data sets are Project 2, Project 3 and MAST II [1996]. All three structures belong to the fully reshaping (armoured mass) berm breakwaters. Figure 4.19 shows the conventional graph of wave transmission versus relative freeboard $R_c/H_{m0,i}$.

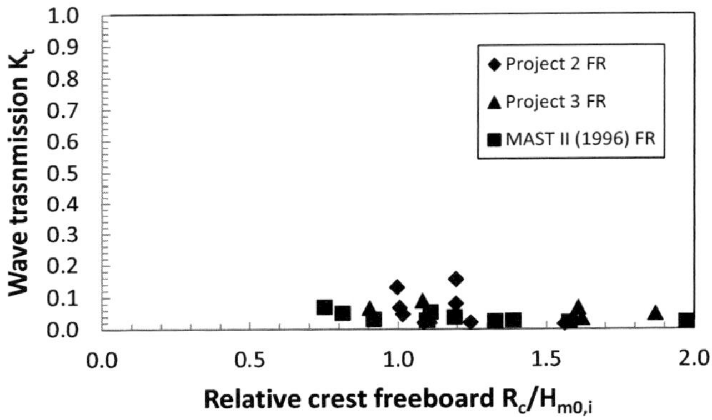

Figure 4.19. Wave transmission for fully reshaping berm breakwaters.

Berm breakwaters are never built as low-crested structures, which means that wave transmission will always be quite small. This is also given in Figure 4.19 as the majority of the data points have wave transmission coefficients less than 0.1. Some of the wave transmission was generated through the crest and some just by overtopping waves directly into the water. Large scale tests in the Delta flume of Deltares showed that there were no noticeable scale effects between a scale of 1:7 and a scale of 1:35. Figure 4.20 shows wave breaking on the reshaped berm of a berm breakwater in the Delta flume, which eventually will give a transmitted wave height behind the structure.

Figure 4.20. A wave breaking on a berm breakwater in the Delta flume of Deltares, giving a transmitted wave height after overtopping.

Chapter 5

Geometrical Design of the Cross-section

5.1 About design guidance in this chapter

Chapter 3 describes the recession at the berm of a berm breakwater for assumed design conditions and main armour rock class (Equation 3.19). It also gives guidance on how some geometrical aspects may influence the recession, (see Section 3.7.4). A flatter lower slope, a higher berm level or a longer berm and a higher toe level all decrease the expected recession. Chapter 4 describes the functional behaviour (wave overtopping, reflection and transmission) and often allowable overtopping rates determine the crest height of the structure.

Chapter 3 and Chapter 4 in principal give expected recession and expected wave overtopping (or requested crest height). Expected recession and crest height are only two parameters that influence the design of the cross-section of a berm breakwater. In order to design a complete cross-section many more design decisions have to be taken. In the past clear design rules on determining a complete cross-section were lacking and design was based on the experience with earlier designs (e.g. in Iceland and Norway) or simply on good reasoning and then testing in a hydraulic laboratory.

It is the experience of Sigurdarson over many years of designing and constructing berm breakwaters in Iceland and in other countries that forms the basis of this chapter on geometrical design of the cross-section. Berm breakwaters may differ in many ways, but were always designed with the experience of the earlier designs and with attempts to improve the design. Focus has often been on a design which limits the volumes of rock, but at the same time increases stability (or decreasing recession).

The implicit design experience of Sigurdarson and colleagues has been made more explicit in this chapter, leading to guidance on geometrical design of the complete cross-section. It is guidance on main principles and every designer has the freedom to deviate from the given principals.

The development of mass-armoured berm breakwaters, often fully reshaping as in the 1980s, to more stable Icelandic-type berm breakwaters, caused a need for guidance on cross-section design. The mass-armoured and the Icelandic-type berm breakwater are therefore treated together in this chapter.

Some of the guidance on geometrical design in this chapter has been published in conference proceedings, like Van der Meer and Sigurdarson [2014] and a practical application in Sigurdarson *et al.* [2014].

5.2 Parameters in geometrical design of the cross-section

5.2.1 *General description of the cross-section*

Principle cross-sections of the mass-armoured and Icelandic-type berm breakwaters were given earlier in Figure 1.1 and Figure 1.9, repeated here as Figure 5.1 and Figure 5.2. These figures show the extremes, a fully reshaping berm breakwater with a wide grading of rock (Figure 5.1) and a hardly reshaping Icelandic-type berm breakwater with four classes of large rock (Figure 5.2). Designs are also possible with three classes of large rock. Mass-armoured types are mainly partly and fully reshaping, where the Icelandic-type gives only hardly or partly reshaping. The main differences between the two types have been described in Section 2.3, where a new classification has been given based on type of berm breakwater and structural behaviour (hardly, partly or fully reshaping). This has led to four distinct classes, each with their range of stability numbers, expected damage and/or expected recession (see Section 2.3).

In both Figure 5.1 and Figure 5.2 similar geometrical parameters have to be established, where for the Icelandic-type, a few more parameters are needed. Figure 5.3 shows the geometrical design parameters of an Icelandic-type of berm breakwater with four classes of rock. A berm breakwater has an upper slope and a lower slope, with in between a berm at a certain level, d_b, with respect to the design water level DWL, and with a certain width, B.

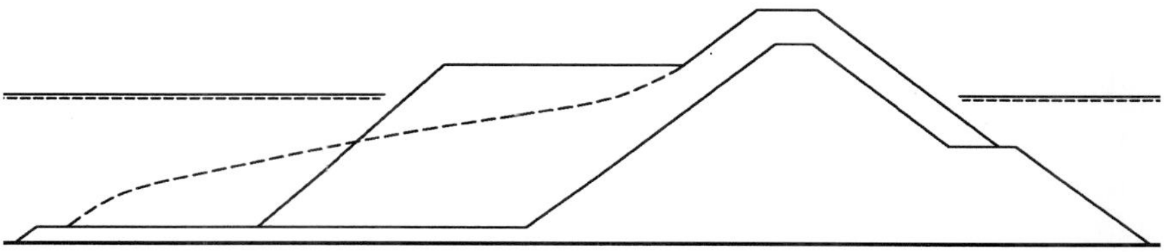

Figure 5.1. Principal sketch of a mass-armoured berm breakwater.

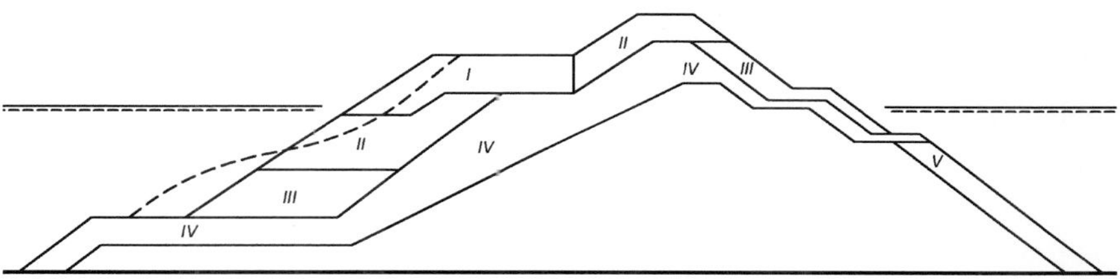

Figure 5.2. Principal sketch of an Icelandic-type berm breakwater.

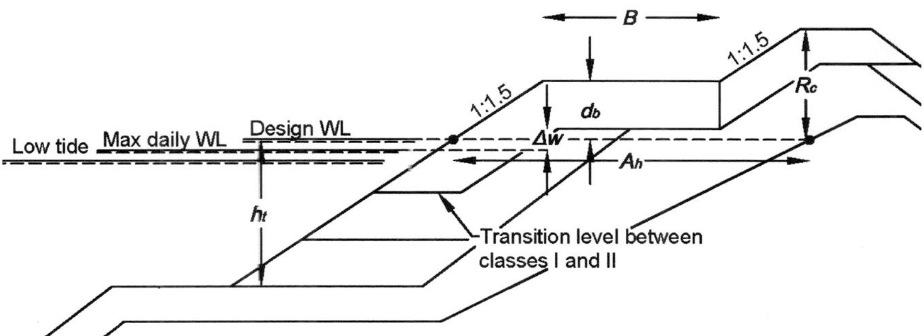

Figure 5.3. Principal cross-section of an Icelandic-type berm breakwater with the main geometrical design parameters.

For a first design of the cross-section, the lower slope as well as the upper slope are taken as $\cot\alpha = 1.5$. Only for the mass-armoured breakwater, a lower slope of $\cot\alpha = 1.2$ is taken as often these type of berm breakwaters have a steeper lower slope. After this first design, the slopes may be modified, depending on actual situations.

The crest height, R_c, should be designed to a certain level, depending on allowable wave overtopping at the design water level DWL. The level of a toe berm (if any) or apron is given by the water depth, h_t, above it. Also the total volume of large rock (Classes I to III or IV for an Icelandic-type) should be sufficient, leading to a "horizontal armour width, A_h", described later in this chapter. Crest width and toe berm width (if any) are in first instance taken as 1 H_{sD}. Depending on actual predicted recession or demands from required space behind the crest, these values may be changed later on.

Tide and daily wave conditions may influence the design of the cross-section, as a safe working level has to be chosen for construction. This level is given as Δw above the daily maximum water level. In the case of an Icelandic-type, this level may influence the final berm level as one may use this level as a working platform (see Figure 5.4). Note that large excavators are able to move on such a layer with large rock (see also Section 7.3).

Figure 5.4. A 110 t excavator working on a level of Class IV rock of 1 to 4 t rock, placing Class II of 10–20 t at the submerged part of the roundhead of the Sirevåg berm breakwater, Norway.

A main parameter is the design water level, DWL, combined with the design wave conditions. The design wave conditions and rock availability may lead to a certain classification of structure (see Section 2.3) to be designed. Structural (recession) and functional behaviour (wave overtopping) may be assessed for lower conditions as well as for higher or overload conditions, but the main design guidance given here on determining the cross-section are based on the 100-years wave height. The maximum water level during this design event is taken as the design water level, DWL, and may influence the choice of the berm level.

A good performance-based design, however, should also look at conditions beyond the 100-years design conditions. These overload conditions, sometimes given as a certain percentage more than the design conditions, or given as a condition for a much larger return period, should be related to the wanted performance. For overload conditions one may accept more recession on the berm and may accept larger wave overtopping, but these conditions should not lead to failure of (parts of) the structure. Even under overload conditions, there should be some part of the berm left and wave overtopping may not lead to severe damage or failure of the crest and or rear of the structure. The design rules developed in this chapter for berm size and crest height will therefore also be applied in Chapter 8 for overload conditions and acceptable recession and overtopping damage.

5.2.2 *Berm width, B, as function of recession and resiliency*

Depending on the classification, the expected structural behaviour is given by more or less recession: hardly, partly or fully reshaping. This has influence on the choice of berm width. It is clear that the lower the stability number is, the more stable the structure will be, with less recession. But less recession should not be the only objective of designing the berm width. The berm width should be (much) larger than expected small recession. With smaller recession, there is a larger capability to cope with extremes, called the resiliency of the structure. This resiliency should play a role in designing the berm width, although it has never been formulated explicitly in design rules.

The berm width is determined from both the need for resiliency and from geometry. The resiliency of the berm breakwater decreases with increasing stability number. One could say that a hardly reshaping Icelandic-type berm breakwater is very resilient to extremes, a partly reshaping mass-armoured or Icelandic-type will show good resiliency and a fully reshaping berm breakwater has marginal or minimum resiliency. It is proposed to consider the following guidelines on resiliency, given in Equations 5.1 to 5.3, connecting expected recession, Rec, and required berm width, B. The resiliency is given as a percentage, $P_{\%}$, of the berm width that may erode under the design condition H_{sD}. For a fully reshaping structure, the percentage should not be 100% as the percentage is given for the design conditions only. Overload conditions larger than the 100-years condition may take more from the berm and for these type of structures one should also consider possible maintenance aspects.

Very resilient, hardly reshaping, IC HR $P_{\%} = 10–20\%$ 5.1
Good resiliency, partly reshaping, IC PR or MA PR $P_{\%} = 20–40\%$ 5.2
Minimum resiliency, fully reshaping, MA FR $P_{\%} \leq 70\%$ 5.3

The reshaping class (hardly, partly or fully) depends on the rock availability. The choice of $P_{\%}$ has to be made by designer and client, where a lower limit can be taken for more safety, but also for a more costly berm. The values given in Equations 5.1 to 5.3 are best guesses based on experience. After having more experience with designing according to these initial guidelines, they should be evaluated and possibly modified. For a good performance-based design it is also necessary to look at the behaviour under overload conditions. For example, for a fully reshaping berm breakwater under overload conditions, one should still want to have P% smaller than 85% or 90%, in order to prevent that recession may end up into the upper slope of the structure.

If the wanted resiliency has been chosen, the berm width follows from:

$$B = Rec/(P_{\%}/100) \qquad 5.4$$

For example, if the expected recession under the design wave height is expected to be 4 m and the wanted resiliency has been chosen with $P_{\%} = 30\%$ (a partly reshaping berm breakwater), then the berm width becomes B = 4/0.3 = 13.3 m.

There is a geometrical criterion for the berm width. As Class I rock of an Icelandic-type of berm breakwater may be quite large, the berm width is very often close to 3 to 4 D_{n50}. A rule for a minimum berm width could be:

$$B_{min} = Rec + 1\ D_{n50} \quad \text{with a minimum of at least } 3\ D_{n50} \qquad 5.5$$

5.2.3 *Crest level, R_c*

Severe wave overtopping may damage the rear side of a breakwater and if this rear is not protected well, the classical failure of a breakwater may occur, i.e. lowering of the structure to about mean sea level. Such severe wave overtopping may occur under overload situations, passing the 100-years condition. It is also for this reason that allowable wave overtopping for the design wave height should not be taken too high. If the rear is not protected by large rock, wave overtopping discharges between q = 10-30 l/s per m may easily destroy the crest of the breakwater.

Figure 5.5 shows results after testing a fully reshaping berm breakwater in Deltares' large Delta flume (scale 1:7) and in a small scale flume (1:35). The small scale tests showed moderate to severe damage at crest and rear, where the large scale test showed start of damage. Results such as these in the small scale tests are not wanted in real situations. Figure 5.6 shows a picture of the testing in the Delta flume, where Figure 5.7 shows a partly reshaping berm breakwater tested under three-dimensional conditions in a wave basin, with severe wave overtopping and resulting damage at the rear of the structure.

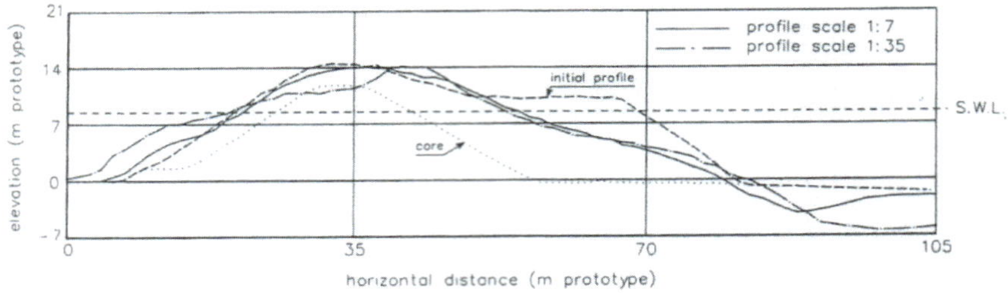

Figure 5.5. Resulting profiles of tests on a fully reshaping berm breakwater. From Van der Meer and Veldman [1992].

Figure 5.6. Testing a fully reshaping berm breakwater in the Deltaflume (Courtesy Deltares).

Figure 5.7. Testing a partly reshaping berm breakwater with severe overtopping (left) and start of damage to the rear (right).

The crest level, depending on the crest freeboard, R_c, can be determined by the new wave overtopping formulae (Equations 4.8, 4.10 and 4.13) if allowable overtopping has been established. Besides that a check can be made, based on structures designed and tested in the past. This design data revealed that most berm breakwater structures have a relative freeboard of:

$$R_c/H_{sD} = 1.0 - 1.2 \qquad 5.6$$

Equation 5.6 could be used for comparison and if no allowable overtopping has been given.

An easy method to check the stability of the rear side is given by Van der Meer and Veldman [1992]. It only applies to the situation as given in Figure 5.1, a fully reshaping mass-armoured berm breakwater and with similar rock in the berm and over the crest and rear. The method gives a stability number, including significant wave height, crest freeboard and wave steepness (using the peak period).

$$R_c/H_s * s_{op}^{1/3} = A \qquad 5.7$$

with A = 0.25 for start of damage; A = 0.21 for moderate damage and A = 0.17 for severe damage.

For the design wave height (100-years condition) one should stay below start of damage (say A = 0.25–0.30), where for the overload situation between start of damage and moderate damage could be taken as the maximum allowable situation (A = 0.21–0.25 in Equation 5.7).

The work of Lykke Andersen [2006] describes rear side stability for other situations with for example very wide berms or crests and partly reshaping of large berms. The crest width and further details of the rear side can be designed as for a conventional structure (see the Rock Manual [2007] or the Coastal Engineering Manual [2006]).

The crest width of a berm breakwater may be limited to a certain number of nominal diameters of the rock that is present as cover layer on the crest. A good design choice for a minimum crest width is:

$$\text{Crest width} = 4\ D_{n50}\ (\text{crest rock}) \qquad 5.8$$

5.2.4 *Horizontal armour width, A_h*

An important design parameter in Figure 5.3 is the "horizontal armour width, A_h". It is the horizontal distance at design water level from the seaward slope of the armour to the transition of sorted rock class to the core. The good structural behaviour of berm breakwaters is, for a large part, due to the large capacity of dissipating wave energy in the large berm. A class of sorted rock gives large voids between the stones and this causes the dissipating capacity. For this reason the horizontal width of the armour at design water level should not become too small. It has always been an implicit but important parameter in design and development of the Icelandic-type berm breakwater in projects in Iceland and Norway.

After design and construction of a large number of these structures, design data were retrieved and an overall summary was derived of this horizontal armour width parameter, A_h, versus the stability number $H_{sD}/\Delta D_{n50}$ (see Figure 5.8). This stability number was chosen as it also gives the division in structural behaviour of berm breakwaters, like hardly, partly and fully reshaping.

At first instance the data points look like a cloud of points, which is mainly caused by the fact that only in recent years the effect of the horizontal armour width on recession became more clear. Therefore, the line in Figure 5.8 was mainly based on the years 2001–2008 and partly on 1996–2000. The linearly increasing relationship becomes more clear.

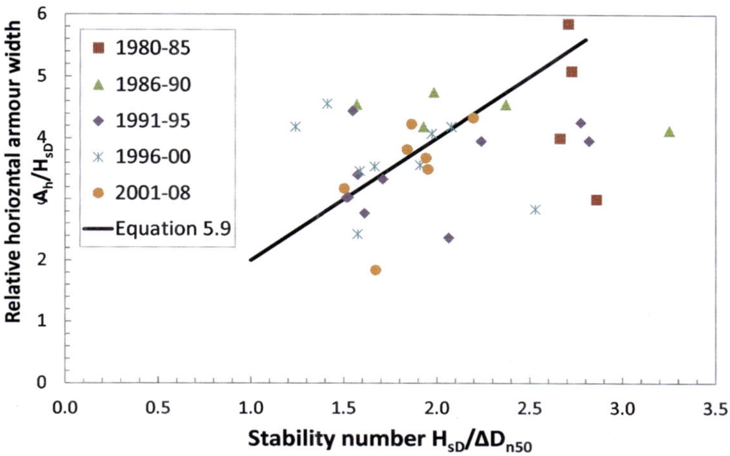

Figure 5.8. Horizontal armour width, A_h, in designs of berm breakwaters over the years.

This analysis leads to a design rule for the horizontal armour width parameter, A_h, which is valid for hardly and partly reshaping berm breakwaters, but also for fully reshaping mass-armoured berm breakwaters:

$$A_h/H_{sD} = 2\ H_{sD}/\Delta D_{n50} \qquad 5.9$$

The horizontal armour width A_h depends linearly on the stability number. Hardly reshaping structures would give A_h = 3.4 to 4.0 H_{sD}, partly reshaping structures to A_h = 4.0 to 5.0 H_{sD} and fully reshaping berm breakwaters to A_h = 5.0 to 6.0 H_{sD}, taking into account the classification given in Table 2.2.

5.2.5 *Rock classes and proposal for new mass-armoured berm breakwater*

The original mass-armoured berm breakwater has a large berm with one rock class, which often has a fairly wide grading, like 1-9 t for example. Figure 5.1 gives the principle cross-section of a mass-armoured breakwater. A mass-armoured berm breakwater is expected to reshape and therefore a lower slope of 1:1.2 can be chosen, which is often easier to construct, while the upper slope should be taken as 1:1.5. The berm level, d_b, is actually free of choice. Sometimes the berm level is just a little above the design water level, and sometimes it is much higher. In principal, the volume of the berm should remain the same, which leads to a long and low berm, or to a shorter and higher berm, but after reshaping they become more or less similar. One should realise that in construction it is often easier to make a berm higher than longer, as a longer berm needs a longer reach of the excavator. Moreover, a higher berm dissipates more wave energy, which may lead to a little less recession (see also Table 3.5).

The main differences between mass-armoured and Icelandic-type berm breakwaters are the design of the berm and expected structural behaviour, i.e. expected recession. The berm of an Icelandic-type berm breakwater is divided in three to four classes of rock, where a mass-armoured berm breakwater has only one class of rock in the berm. The design of an Icelandic-type berm breakwater is based on limiting recession by putting the large rock where it is needed, often using a very large rock in a small area and making this as stable as possible.

The idea of using large rock where it is needed, was explored in the European research project MAST II, [1997]: Berm breakwater structures, (MAST-Contract MAS2-CT94-0087). Figure 5.9 shows the set-up of four tests with different use of rock in the berm.

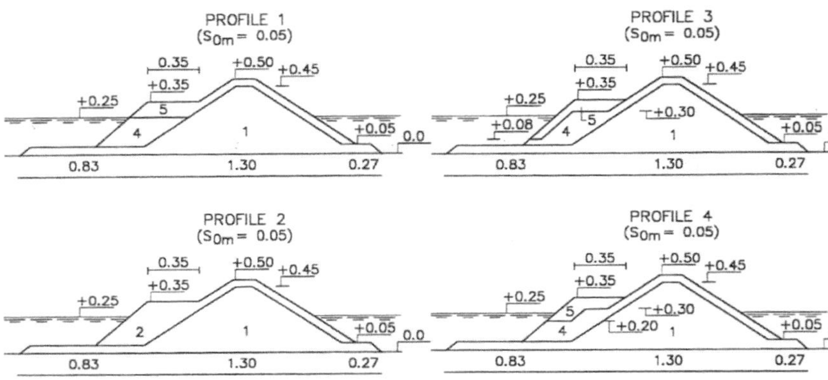

Figure 5.9. Set-up of tests where the berm rock of 1–12 t in profile 2 was divided in two classes with 1–6 t (Class 4) and 6–12 t (Class 5). MAST II [1997], using a linear scale factor of 50, Juhl and Sloth [1998].

Note that the graph was copied from Juhl and Sloth, [1998] and not all information shown may be useful here. Also large rock classes were numbered 2, 4 and 5, where we use Class I for the largest armourstone and subsequently Classes II and III for smaller stone classes.

Profile 2 gives the conventional mass-armoured berm breakwater with only one class of rock. For profiles 1, 3 and 5 the available rock in the berm of profile 2 was divided in two classes, where Class 5 was the larger class. The small rock Class 4 was in volume about to times larger than rock Class 5 (divided in two-thirds and one-third in volume).

The testing was basic research without a linear scale factor to real situations, therefore measures in Figure 5.9 are given in m model scale. By using a linear scale factor of 50 the tests can be compared to real situations. The conventional berm breakwater, profile 2, had a rock class equivalent to 1-12 t with a grading $D_{85}/D_{15} = 1.80$. This class was divided in stone Class 4 with 1–6 t (grading $D_{85}/D_{15} = 1.65$) and stone Class 5 with 6–12 t ($D_{85}/D_{15} = 1.20$). The mean weights of the three rock classes 2, 4 and 5 where respectively $M_{50} = 3.8$ t, 2.6 t and 9.5 t.

The results of the testing are shown in Figure 5.10 where actual recession, Rec, is given versus the wave height H_{m0}. It is clear that putting larger rock on seaward face and upper berm, with smaller rock underneath, gives less reshaping than a large berm with one wide grading of rock. Note that the volume and size of rock is exactly the same in all tested cross-sections, only the location of sizes of rock is different.

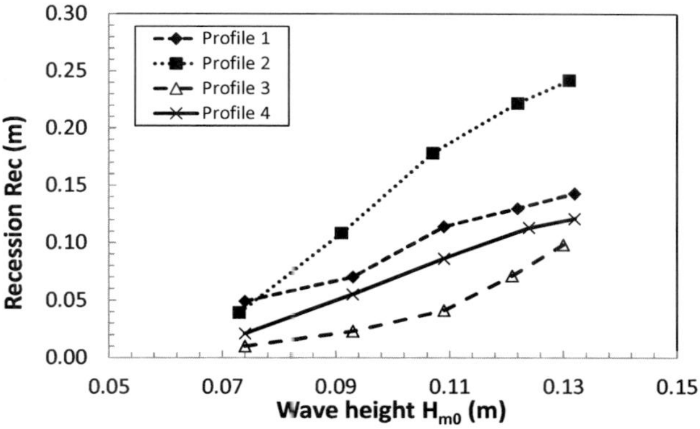

Figure 5.10. Results on recession for profiles in Figure 5.9. Graph re-drawn from [MAST II, 1997], [Juhl and Sloth, 1998].

The (predictable) knowledge as described above, that larger rock would give less recession, was used to develop the Icelandic-type berm breakwater from the original mass-armoured berm breakwater. Often large rock was used anyway and the rock had to be placed individually. In such a case, costs do not really increase if more rock classes are used, handled and placed than for the mass-armoured case.

Figure 5.3 shows the principal cross-section of an Icelandic-type berm breakwater. In this case four main armour rock classes I–IV are given, but for smaller structures this could also be reduced to three. The main class is Class I on upper seaward front and on the berm. The principal slopes are 1:1.5 on seaward lower and upper slope. In the present case the slope of the core is taken as 1:2 in order to reduce some volume of large rock. But a 1:1.5 slope could also be applied if that would be easier from construction point view. Then, the slope of the core should be rotated around the right point of the horizontal armour width, A_h.

An important geometrical aspect to consider is the division in several rock classes. Class I will be the largest rock class on berm and upper seaward slope. If this one has been determined, consecutive smaller classes have to be chosen. This section gives some principles and examples.

There is a significant difference if standard gradings are used [Rock Manual, 2007], or if a dedicated quarry has been or can be opened. In the case of rock supply from existing quarries with standard gradings, the design conditions together with the standard gradings give the stability number that can be achieved and the type of Icelandic berm breakwater (hardly or partly reshaping). The standard heavy gradings as given in the Rock Manual (2007) are given as: 10–5 t; 6–10 t; 3–6 t; 1–3 t; and 0.3–1 t. Depending on the maximum grading to be used (= Class I), one can use the lighter gradings for the other classes.

In the case of a dedicated quarry there are various aspects to consider to come to an optimised division in rock classes. The *in-situ* block size distribution may be used to derive a quarry yield prediction. The maximum size, M_{max}, of this prediction plays an important role in choosing the Class I rock. But subsequent gradings depend on the required volume of that grading for the breakwater as a whole and the percentage that is predicted from the quarry yield. It may then be an iterative process to derive rock gradings that match the predicted quarry yield as close as possible. Table 5.1 gives a few examples of possible gradings. In this table the maximum grading has been given as rock exceeding at least 15 t. It is possible that such a large grading cannot be achieved, or that it is simply not required for relatively low design wave heights. In such a case Class II can be promoted to Class I level.

Table 5.1. Examples of gradings from dedicated quarries.

Class	*Example 1*	*Example 2*	*Example 3*
I	15-30 t	16-30 t	20-35 t
II	6-15t	10-16 t	10-20 t
III	2-6 t	4-10 t	4-10 t
IV	0.5-2 t	1-4 t	1-4 t
V		0.3-1 t	

The original fully reshaping berm breakwater was designed with only one armour class with a wide grading. The Icelandic-type berm breakwater has often three or four large rock classes, as described above. It is fairly easy to divide a wide rock class into two rock classes with narrower gradations. It is common practice for Icelandic-type of breakwaters, but not really for fully reshaping berm breakwaters. By developing the geometrical design rules in this book, it became clearer and clearer that a large and fully reshaping berm is actually neglecting the fact that the largest rock is only needed where the wave action is. By using a wide grading for the full berm, a lot of fairly large rock is wasted down at the bottom and well within the berm, without a clear purpose. The original idea was that a berm with only one rock class would be easier to construct. Practice has proven over the years that costs do not increase if a large berm is divided into more rock classes. It is for this reason that a new proposal is brought forward for a "mass-armoured" berm breakwater. The idea is given in Figure 5.11.

In the original design a berm would consist of, for example, a wide grading of 1–9 t rock. It is relatively easy to divide this class into two gradings of 1–4 t and 4–9 t. Then, the smallest rock is used for the lower part of the berm and the larger rock at the upper part.

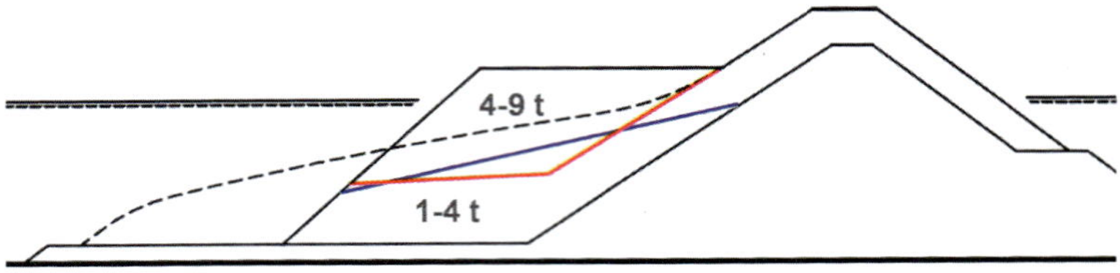

Figure 5.11. Proposal for a new "mass-armoured" berm breakwater: two more narrow graded rock classes instead of one wide grading.

It is the designer's decision to choose between the configuration with the red or with the straight blue line. Both options must be seen as comparable with respect to reshaping. With this modification the design moves a little towards an Icelandic-type of berm breakwater, but still has no rock layers, just two volumes of rock: a lower part and an upper part. Construction costs of a design like Figure 5.11 will be comparable to the original design with one wide grading and exactly the same rock is used. But the design of Figure 5.11 is more stable, giving less recession, as the stability number increased due to a larger D_{n50} for the 4–9 t rock than for the 1–9 t rock.

If easy construction is wanted or required, a mass-armoured berm breakwater may be an option. But it is strongly proposed to decrease recession and increase resiliency, without increasing costs, to use two classes of rock as given as an example in Figure 5.11.

5.2.6 *Berm level, d_b*

A mass-armoured berm breakwater needs a certain volume of rock in the berm to cope with expected recession and reshaping. A low berm, closer to the design water level, will result in a wider berm. The mass balance should be more or less fulfilled, which is the basis for a modification of berm level. This means that the cross sectional area above the berm could be replaced by a similar area in front of the original berm or vice versa.

Another modification might be a flatter lower slope than 1:1.2 in order to limit the initial reshaping. The rotation point for the slope will be half way the level of the foundation layer and the top level of the berm.

The procedure is similar for a steeper slope, say up to 1:1, when berm rock is simply dumped without creating a flatter slope (angle of repose). The first design of fully reshaping berm breakwaters in the eighties had a steep slope as the storm waves were expected to create a stable berm. For fully reshaping berm breakwaters, the initial berm level is not a real issue, more the total volume of rock, although a higher berm level will probably give a little less reshaping.

The berm width B as defined in Figure 5.3 is given by wanted resiliency, described in Section 5.2.2 and Equations 5.1 to 5.4. The berm level d_b often depends on two aspects. The first and easiest is that the berm level should be at least 0.6 H_{sD} above design water level. Such a high berm increases stability, see also Section 3.7.4. Of course the berm may be designed at a lower level, but one should then expect more recession.

$$d_b \geq 0.6\ H_{sD} \qquad 5.10$$

The other aspect on berm level might be the level for construction work. Often the top level of the Class II, III and IV, which gives the transition to Class I rock at the berm determines with 2 D_{n50} of Class I rock on top the final level of the berm, see Figure 5.3. This level is recommended as a working level for the excavators placing the Class I rock (see also Figure 5.4).

The minimum working level depends on the local situation with tides and daily waves and assessment of such a level would not differ much from procedures given in the Rock Manual [2007] or Coastal Engineering Manual [2006]. For example, one could take mean high water spring (MHWS) with extra safety for daily waves. This extra safety, given as Δw in Figure 5.3, depends on the actual wave climate expected during construction to ensure safe working conditions, see Figure 5.3. For low wave activity Δw has often been chosen between 0.5 m and 1 m and for more moderate wave activity between 1 m and 1.5 m.

In conclusion, the lower level of the two diameters thick Class I is based on Equation 5.10 or on local tide levels (for example mean high water spring) and a safe level for daily wave conditions. Then, two diameters D_{n50} of the Class I rock have to be added to come to the final berm level.

5.2.7 *Apron*

Figure 5.1 and Figure 5.11 show that the large rock of a fully reshaping berm breakwater is laid on a foundation layer with an apron in front of the berm. Such a layer is really required if the structure is founded on sand or other movable material (and not on rock). This is a common rule for conventional breakwaters, where a geometrical tight, or sometimes geometrically open, filter layer has to be designed, or where a geotextile is used. Also berm breakwaters need such a foundation layer, where sand cannot escape from beneath the structure. For a reshaping berm breakwater, large rock from the berm will fall down the slope and form a longer S-shaped profile, as given in Figure 5.1 and Figure 5.11. Therefore an apron or foundation layer has to be designed in front of the berm to provide a foundation for the reshaping berm.

The consequences of not providing a foundation layer and apron are shown in Figure 5.5 and Figure 5.12. The tested structures were built on a sandy foreshore, where the large rock was placed directly on the sand. The effect is that a small scour hole is developed in front of the berm. Then rock in the berm subsided directly into the sand and the scour hole, losing rock from the profile, leading in turn to heavier wave overtopping. Figure 5.12 shows schematically the final results of the subsidence when the structures was dismantled after testing. Further testing with a foundation layer showed none of this subsidence.

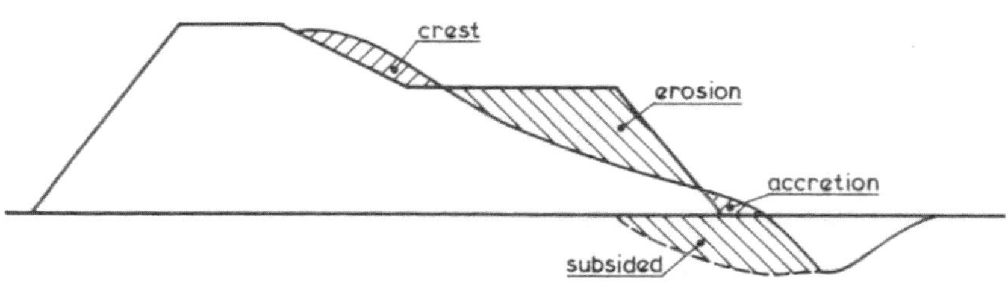

Figure 5.12. Subsidence of large rock from the berm into sand if no foundation layer has been designed. From Van der Meer and Veldman [1992].

An apron is also required for an Icelandic-type of berm breakwater, but this can be much shorter as the recession will be less for a hardly or partly reshaping structure.

5.2.8 *Transition from Class I to Class II rock*

The primary goal of the tests performed by Sveinbjörnsson [2008] was to find design guidance about the depth on the lower slope of the transition of Class I to Class II rock. He tested several configurations of Icelandic-type berm breakwaters with different stone classes and transitions at various levels. His final guidance gives:

> Class I stones are recommended to reach below LAT as far down as: $h_{I\text{-}II} \geq 1.45 \cdot \Delta D_{n50}$, Class I or $h_{I\text{-}II} \geq 1.85 \cdot \Delta D_{n50}$, Class II, where $h_{I\text{-}II}$ is the transition from Class I to Class II rock on the lower slope. The one of the two that gives the larger $h_{I\text{-}II}$ in each case is the recommended choice.

By taking $\Delta = 1.6 - 1.7$ the transition would be about 2.5 D_{n50} Class I below the water level considered and this would be regardless of the stability number used for design. The hardly and partly reshaping Icelandic berm breakwaters are especially designed with fairly low stability numbers of $H_{sD}/\Delta D_{n50} = 1.7 - 2.5$, where recession is limited and Class I rock is fairly large. Then a transition at 2.5 D_{n50} below the water level is considered a long distance and longer than what is used at most structures designed and constructed in the past. Based on earlier designed structures it is proposed to use 0.4 H_{sD} as a limit.

Using this 0.4 H_{sD}-limit should also consider lower water levels with adjusted design wave heights. Structural behaviour (recession) and wave overtopping are often considered at the design water level, DWL, which is the highest water level that can occur under design conditions (100-years event). But it is also possible that the design wave height is present for some time at lower levels, for example during low tide, but including surge. For relatively deep situations the wave height will remain more or less the same for lower water levels, but in depth-limited situations the wave height will decrease with lower water levels considered. The 0.4 H_{sD}-limit for the transition from Class I to Class II rock has to be measured from the lowest water level that can occur with more or less the design conditions (within 90% of H_{sD}).

For a mass-armoured berm breakwater, with two classes of rock as proposed in Section 5.2.5 and Figure 5.11, a good transition might be 0.6 H_s below the considered water level.

5.2.9 *Possible toe berm*

Figure 5.3 also shows a foundation level for the large rock classes of an Icelandic-type berm breakwater, well above the level of the foreshore. In the graph this is given as Class IV on a horizontal layer. For relatively deep water as well as for depth-limited conditions it is favourable for the stability of the berm if this foundation level is as high as possible, see Table 3.5. In principle one could consider this structure as a toe berm structure for a conventional breakwater and use toe stability formulae to assess the stability. One such formula is given amongst others in the Rock Manual [2007, Equation 5.188]:

$$\frac{H_s}{\Delta D_{n50}} = \left\{2 + 6.2\left(\frac{h_t}{h}\right)^{2.7}\right\} N_{od}^{0.15} \quad 5.11$$

where h_t = the toe depth and h the water depth in front of the toe, both related to the water level considered and D_{n50} is calculated for the toe armour. N_{od} is a damage level and for the kind of structure as in Figure 5.3 for a berm breakwater, an allowable damage level of N_{od} = 2 for the design conditions is a good choice. Physical model testing may show that even a larger damage could be acceptable (a little reshaping of the corner). For more information one is referred to the Rock Manual (2007). The water level to be applied in Equation 5.11 is the maximum design water level, DWL, with the design wave height, H_{sD}, as well as a low water level with according wave height, based on the 100-years event.

There is more guidance on toe berm stability of rubble mound and caisson breakwaters, but they are not always comparable. A recently published alternative is the formula on toe rock stability of Van Gent and Van der Werf [2014]. Here we use Equation 5.11, but it is the freedom of the designer to use and compare with alternative prediction methods of toe berm stability.

A check should be made whether the level of the designed toe can indeed be constructed (see also Figure 5.3). It is assumed that the core extends seaward with at least a thickness of 1.5 m. Then on top of this core the rock layer of the toe berm will be constructed, which has a thickness of at least 2 D_{n50}. The lowest level of the toe berm is then 1.5 m + 2 D_{n50} above the foreshore.

5.3 Application of geometrical rules for a design

It sometimes happens that recent developments may be applied directly in a real design. This occurred when the draft geometrical guidance in this chapter was developed and could almost directly be applied to an alternative design for a berm breakwater coastal defence of a reclamation area at Hambantota in Sri Lanka. Background information on this project can be found in Sigurdarson *et al.* [2014].

It was assumed that Class I rock of 5–10 t could be achieved from quarrying operation. The design wave height for swell waves was H_{sD} = 5.6 m with a wave period of T_p = 18 s and a possibility of cyclone waves with a wave height of H_{sD} = 6.5 m, but with a shorter period of T_p = 12.5 s. The water depth was 14 m.

The geometrical design parameters were calculated according to the guidelines given in this chapter and have been summarized in Table 5.2. The breakwater design should be considered as fully reshaping as the stability number was between $H_s/\Delta D_{n50}$ = 2.5 - 3.0. Quite some reshaping could be expected according to Equations 3.19 and 3.20. Some extra safety was taken into account as the experience of the contractor with berm breakwaters was very limited. This resulted in a resiliency of only P = 40%, leading to a large berm width of 18 m.

Based on the geometrical design guidelines, a spreadsheet was developed, which finally gave a first indication of a cross-section. The outcome for the Hambantota case is given in Figure 5.13. It shows the expected recessions for swell as well as cyclone waves and also for the overload conditions. The graph also gives an indication of three rock classes to be applied, where the layer thickness of Class I on the seaward side was 4 D_{n50}. The horizontal armour width, A_h, is also given in the graph.

Table 5.2. Geometrical design considering swell waves; Hambantota, Sri Lanka.

Design conditions	*Parameter*	*Value*
Design wave height	H_{sD} (m)	5.6
Peak period	T_p (s)	18.0
Overload wave height	H_s (m)	6.2
Design water level, DWL	(m CD)	1.0
Allowable overtopping q for H_{sD}	(l/s per m)	10
Allowable overtopping q for H_{sOL}	(l/s per m)	30
Selected stone classes		
Class I	(t)	5-10
Stability number	$H_s/\Delta D_{n50}$	2.6
Type of berm breakwater		FR
Length parameters		
Recession for H_{sD}	Rec (m)	5.2
Recession for H_{sOL}	Rec (m)	6.4
Wanted resiliency		40%
Min. berm width from resiliency	B (m)	12.9
Min. berm width from geometry	B (m)	6.5
Min. berm level from waves	d_b (m CD)	4.4
Min. berm level from constructional issue	d_b (m CD)	5.8
Min. horizontal armour width	A_h (m)	29.1
Min. transition level between Classes I and II	h_{I-II} (m)	2.2
Crest level and crest width		
Overtopping influence factor	γ_{BB}	0.60
Required crest level for H_{sD}	R_c+DWL (m CD)	10.8
Required crest level for H_{sOL}	R_c+DWL (m CD)	10.4
Required crest width	(m)	5.6

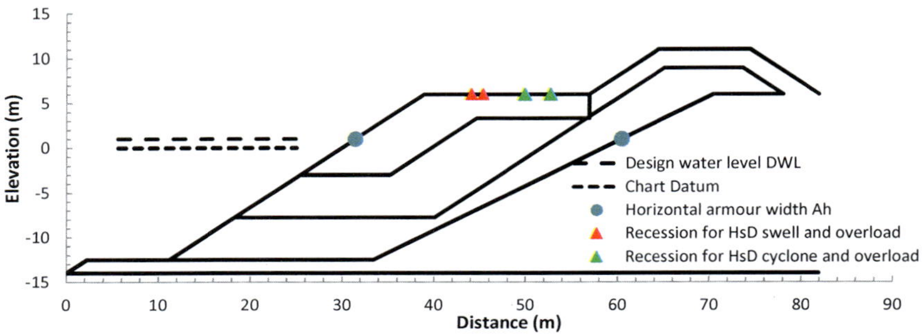

Figure 5.13. Outcome of spreadsheet with applying the geometrical design guidance for the breakwater of Hambantota.

Note that the recession was based on Equation 3.19 and Equation 3.20 only and that positive effects, as given in Table 3.5, were not taken into account at this stage to make an adjusted calculation for the recession. Positive effects were mainly a high and long berm and a relatively gentle slope as well as an elevated toe berm (not in Figure 5.13).

Model testing was carried out in Tianjin Port Engineering Institute (TPEI) in China. The results, certainly of the first test, were unexpected. Figure 5.14 gives the results of the first three tests. The reshaped profile was not measured for all wave conditions and the measurements were performed manually, showing some scatter. But the conclusion is clear: the berm was much more stable than according to the average curve by Equation 3.19 and Equation 3.20.

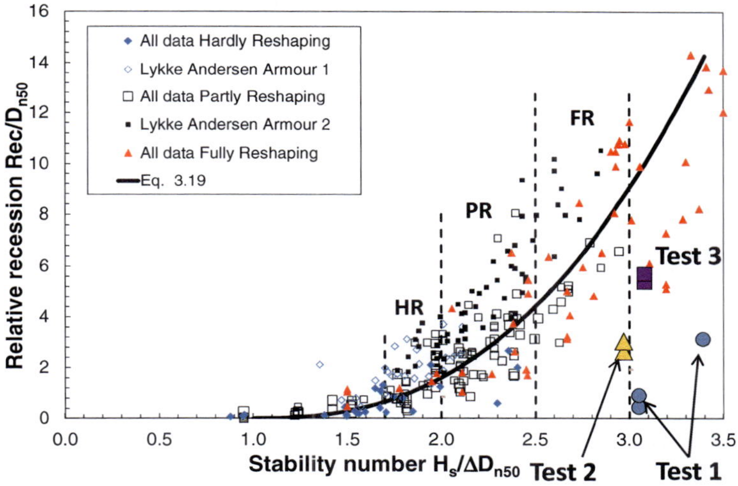

Figure 5.14. Test results of the first three tests of the Hambantota breakwater, showing much less reshaping than the average curve (Equation 3.19).

The results of test 1 were so unexpectedly stable that they were hard to believe. A closer inspection of the rock showed that the rock had been painted with quite sticky paint, which remained sticky after it was dry. Moreover, the shape of the rock was very angular with sharp edges. It was concluded that this test was not according to standard test practice. The rock was tumbled in a laboratory concrete mixer to lose its sticky paint and to lose the very sharp edges. Tests 2 and 3 were performed with this rock.

In test 2, a toe berm was added at a level of -8 m. This test showed a recession of about 3 rock diameters, which is still much less than the expected 8 or 9 diameters, based on Equation 3.19 only. Test 3 was performed with a shorter berm width of only 10 m instead of 18 m. Now the recession increased to almost 6 rock diameters. The results of that test were close to other tests in the data set in Figure 5.14. It was concluded that the large berm width had also a positive influence on recession.

A further analysis was made on the chosen design parameters in combination with Table 3.5. It can be concluded that the initial design, including the addition of a toe berm in tests 2 and 3, could be expected to be more stable due to various reasons:

- The berm level d_b was above 0.6 H_{sD}, which increases stability;
- The berm width was quite longer than necessary, which increases stability (see the difference between tests 2 and 3 in Figure 5.14);
- The lower slope was the most gentle one used for berm breakwaters;
- The toe berm was quite high, which increases stability.

These four reasons were enough to limit the recession to only one-third of the expected recession based on Equation 3.19. At first sight the improvements were taken implicitly, as we were aware of the positive points with regard to recession. At hindsight it can be concluded that the limited recession was indeed due to a combination of four positive points.

The tests resulted in a final design where the berm width was reduced to 15 m, leaving all the other points intact. The final design is given in Figure 5.15.

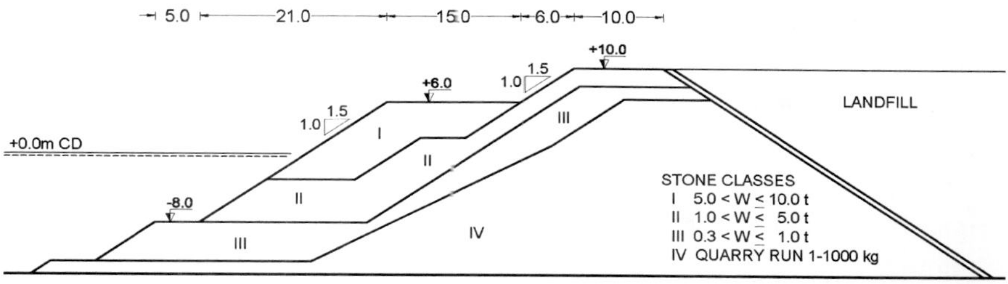

Figure 5.15. Final design of the Icelandic-type berm breakwater at Hambantota, optimised during physical model testing.

5.4 Roundhead

Till now reshaping in this chapter has been considered to come perpendicularly to the alignment, i.e. changing the sectional profile. Oblique wave attack may reduce this recession, certainly if the angle of wave attack becomes larger than say 30°. In most cases recession at a trunk section of a berm breakwater means that rock is falling down onto the lower slope. Only for gentle (reshaped) slopes and rather large stability numbers, some rock may also be transported upwards, forming a kind of rock beach crest.

The situation is different for a roundhead of a berm breakwater, but to some extent similar to conventional breakwaters. Attack on the round head of a breakwater means that wave forces have a direction along and around the head and not only up and down as for a trunk section. For berm breakwaters, this means that rock can be transported backwards to the rear of the round head.

There is a significant difference for roundhead behaviour for fully reshaping berm breakwaters and for the more stable Icelandic-type berm breakwaters. The latter having also limited recession at the roundhead. Basic tests on roundhead behaviour for "berm breakwaters" were performed by Burcharth and Frigaard [1987]. In reality the structures tested were not really berm breakwaters, as the stability numbers for testing were between $H_s/\Delta D_{n50} = 3.5 - 7.1$, well above the criterion for stable berm breakwaters of $H_s/\Delta D_{n50} = 3.0$. Moreover, the cross-section was a homogeneous cross-section with only one grading of rock. It should be noted that these tests were performed at the start of design of modern berm breakwaters, where guidance was simply lacking. Indeed, the occasion for presentation of the work by Burcharth and Frigaard [1987] was the first conference on berm breakwaters.

Figure 5.16 gives the result of one of the tests of Burcharth and Frigaard [1987] for a stability number of $H_s/\Delta D_{n50} = 5.4$. There is larger reshaping mostly above the water line and a large amount of rock has been transported to the rear of the head. Although the figure is not representative of the real reshaping of berm breakwaters, it gives a good idea of the way reshaping occurs.

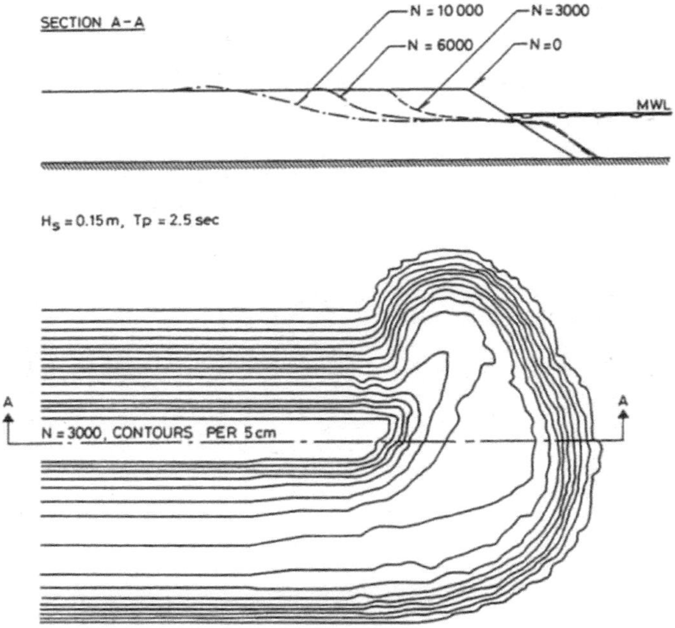

Figure 5.16. Reshaped roundhead of a dynamically stable berm type structure. Note that the stability number for this test was $H_s/\Delta D_{n50} = 5.4$, well above the criterion for a statically stable reshaped berm breakwater. Figure from Burcharth and Frigaard [1987].

An example of testing of a roundhead for a real project was described in Van der Meer and Veldman [1992]. The fully reshaping berm breakwater was situated in depth-limited conditions with a maximum significant wave height in overload conditions (increased water level) of

6.1 m, where the deep water wave height was over 10 m. The rock of the berm consisted of 1.5–9 t rock, which gives a design (overload) stability number of $H_s/\Delta D_{n50} = 3.11$, just over the limit for statically stable berm breakwaters. The plan view in a wave basin with test results is shown in Figure 5.17.

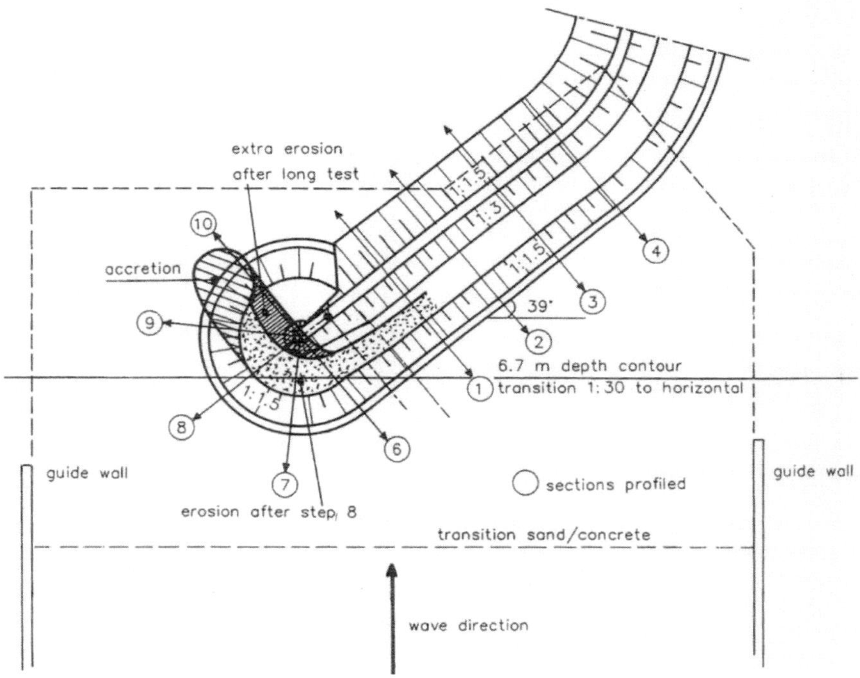

Figure 5.17. Plan view of reshaping of a berm breakwater roundhead with a stability number of $H_s/\Delta D_{n50} = 2.87$, for the normal test duration of 6 hours and for a long test with 36 hours (prototype dimensions). Figure from Van der Meer and Veldman [1992].

The berm elevation of the trunk was 3.6 m above Chart Datum and the design water level DWL was 1.8 m above Chart Datum. The trunk was statically stable after reshaping. In order to have more resistance to wave attack on the head, the elevation of the berm of the head was increased to 6.1 m above Chart Datum, giving a larger volume of rock in the berm. Eight sub-tests of 6 hours each were performed. After this complete test, the conclusion on the stability of the round head was that it performed well. Sections 7–9 in Figure 5.17 show some erosion at the berm around the still water level, but do not show erosion at the crest.

After the normal test procedure it was decided to test the reserve capacity or resiliency of the head under very severe wave loading. The most severe sub-test, described above with a stability number of $H_s/\Delta D_{n50} = 2.87$, was run again, but now for a duration of 36 hours, six times longer than in the normal test procedure. The extra erosion after this condition are also shown in Figure 5.17. The erosion at sections 6–9 increased considerably. Even the core became visible at the crest of the round head. Nevertheless the head did not fail in a catastrophic way and survived the extreme wave loading very well.

Figure 5.17 shows a plan view of the erosion and accretion. It shows the erosion of the berm at the trunk and the head after the normal test procedure as well as the accretion at the inner side of the head at section 10. Besides the transport of material from the berm to the toe of the structure, a part was transported along the wave direction and this caused the accretion at section 10. The severe and long duration test after the normal test procedure showed a considerable increase in erosion at the head. The actual amount of eroded material, however, is in fact rather small as can be seen in Figure 5.17, where this area is given by the dark shaded part. Although the core became visible after that condition, this was only very locally at the tip of the round head.

The overall conclusion on the stability of the round head was that increasing the height of the berm of this fully reshaping berm breakwater and therefore creating a larger amount of rock at the head, can be seen as a good measure for enlarging the stability of the round head of such a berm breakwater, using the same rock size as for the trunk.

Rock for an Icelandic-type berm breakwater is often larger than for a fully reshaping mass-armoured berm breakwater. Part of the developments of an Icelandic-type berm breakwater is always finding the largest feasible rock in a dedicated quarry. Very often less than 5% of the quarry yield is required to construct the berm breakwater with Class I rock at the most attacked locations. The demand for Class I for the berm of the roundhead is even less than 5%. Often only 1% is needed to give a grading that is even a little larger than the Class I rock for the berm of the trunk. In this sense, one should challenge the output of the quarry.

The berm of a roundhead for an Icelandic-type berm breakwater has often the same cross-section as the trunk-section, but with larger rock to reduce recession. It may well be that the berm level increases a little, if the working level is maintained at the same level. The larger rock gives a slightly higher berm in that case. The behaviour of the roundhead with (slightly) larger rock than for the trunk-section is similar to that of the trunk-section, with similar recession. Data on rock size for a number of Icelandic-type berm breakwaters in Iceland and Norway have been analysed and the results are summarised in Figure 5.18. In all cases the structures were constructed from dedicated quarries, where it is easier to get a larger size of rock than a quarry that produces standard gradings. If the maximum standard grading has been used for the trunk, it is difficult to increase further the size of rock for the head.

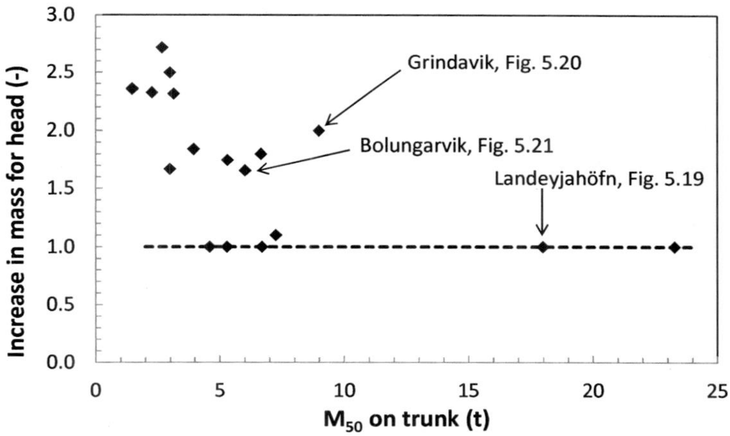

Figure 5.18. Increase of rock size on head compared to the trunk section. Data of constructed Icelandic-type berm breakwaters in Iceland and Norway, all with a dedicated quarry.

Figure 5.18 shows a certain trend. For quite small rock sizes, say M_{50} on the trunk smaller than 5 t, it is quite easy to double the mass for the rock on the head. For rock classes on the trunk that are larger than 10 t, the increase in mass is less than a factor of 2 and often even 1 (no increase). It is quite difficult to find larger rock if the trunk size is already quite large, such as 18 or 23 t in the graph. Although data are missing between 10–18 t in the graph, one may conclude that if the trunk size is between 10–20 t, the increase for the rock on the head will not be much more than a factor between 1.0 and 1.3.

Examples of berm breakwater roundheads for Icelandic-type breakwaters are given in Figure 5.19, Figure 5.20 and Figure 5.21. The breakwater of Landeyjahöfn in Figure 5.19 had already a large Class I of 12–30 t armourstone and it was not possible to increase the weight.

Figure 5.19. Example of an Icelandic-type berm breakwater at Landeyjahöfn with similar rock 12-30 t at trunk and head section.

The narrow navigational channel between the two breakwaters of Grindavik (Figure 5.20) is vulnerable for rock moving from the roundheads. Moreover, the design wave height at the heads was higher than at the trunk. This led to a required increase of Class I armourstone from 6–15 t to 13–28 t, giving a M_{50} increase of a factor of 2. In that specific case, it was not possible to find enough larger rock for the entire head sections. For that reason only the section 90°–160° with respect to the angle of wave attack was strengthened with larger armourstone

13–28 t. This is the section where wave attack on a round head is most severe. The pictures in Figure 5.20 show the larger stones and the slightly increased berm level, for the upper picture on the right hand side and for the lower picture on the left hand side. Although frequently being exposed to breaking waves, not a stone has moved after more than 10 years in service.

Figure 5.20. The roundheads of the two Grindavik berm breakwaters, partly reinforced at the harbour side with 13 to 28 t armourstone.

The Bolungarvik breakwater was built in 1992 and 1993 (see Figure 5.21). It is 300 m long and the deepest part is located in about 9 m water depth CD. Class I armourstone on the trunk section was 4-10 t with M_{50} = 6 t. The head section was protected with 8-14 t armourstone in two layers on top of the berm with M_{50} = 10 t, giving a M_{50} increase of a factor of 1.66.

The design wave consists of swells with a significant wave height of 6.3 m and a peak period of 17s. In January 1995 it was exposed to a storm close to design storm conditions lasting for at two days. Till today the breakwater has only experienced minor reshaping at a few locations not exceeding 2 D_{n50}.

Figure 5.21. The Bolungarvik breakwater.

5.5 Soft soil

When breakwaters are built on soft soil this involves a risk of liquefaction. Several methods exist to deal with those conditions and reduce liquefaction hazards. Often the soft soil is removed by dredging and replaced by stable material, or soil improvement techniques are applied to avoid large increase in pore water pressure. But these methods are rather costly, mobilisation of a dredger and the cost for removing and replacing the soft soil. When local conditions are known, it is often possible to apply cheaper methods like gradually filling material over the soft soil to increase its shear strength by consolidation.

This method has been used in several breakwater projects in Iceland, both with berm breakwaters as well as with conventional rubble mound breakwaters. From 1998 to 1999 a berm breakwater was built as a part of an extension of the Port of Hafnarfjördur in Iceland, Figure 5.22, [Sigurdarson *et al.*, 1999]. The breakwater was built in about 11 m CD water depth with a mean spring tidal elevation of +4 m CD. The seabed consisted of more than 20 m of soft organic silty soil and sandy in the uppermost three meters. The water content of the soil varied between 60% to 80% and it was classified as OL, ML-OH according to the Unified Soil classification System with the following soil characteristics:

- Soil unit weight 17 kN/m^3 in uppermost 3 m, decreases to 15–16 kN/m^3 below 3 m depth;
- Undrained shear strength 15–20 kPa down to 10 m;
- Friction angle under drained conditions is 40° with a cohesion value of 0.

Figure 5.22. The breakwater in the Port of Hafnarfjördur shelters a multipurpose quay and a ship repair basin with floating docks. From the land area the breakwater starts with a northerly direction, then turning towards east, before the end is turned toward north-east. Photo by Mar Sveinbjörnsson.

Stability analysis showed that the filling of the breakwater foundation or sublayers should be done in stages to allow the foundation soil to accommodate increased shear stresses by consolidation. The design included seismic calculations for an earthquake of magnitude 6.0 to 6.3 on the Richter scale, occurring at a distance of 17 to 20 km. Based on evaluation of available CPT data it was concluded that liquefaction of the foundation soil under the centre of the breakwater was unlikely. However, liquefaction was likely near the end in front of the toe of the breakwater. In the final design the breakwater therefore was strengthened by 5 m wide extra berms on both sides.

The filling from sea was done by a hopper dredger. The designers would have liked several months in between the filling stages for the pore pressure to dissipate, but the harbour authority pressed for a shorter construction period and the filling was completed in six months, from March to August 1998. The first filling layers, up to a water depth of -4 m CD, consisted of dredged shell sand, a relatively light material and inexpensive. This was later filled over by coarse basaltic sand and gravel up to an elevation of about 0 to +1 m CD (see Figure 5.23). Before filling from land started, wave erosion and settlement had lowered the mound by about a metre.

Figure 5.23. Filling of the dredged basaltic gravel on the North-South trunk of the Hafnarfjördur breakwater, finished up to about 0 to +1 m CD.

Filling from land of blasted material, quarry run and armourstone, started shortly after filling from sea had finished. It should have been completed in only a year, but construction was extended by several months. The settlement was monitored carefully by measuring on wooden poles that were rammed into the fill mound. The predicted settlement was about 2 m, but mainly due to the shorter settlement time, the total settlement became close to 4 m in some areas (Figure 5.24). This called for revision of the berm breakwater design during the construction. Due to the flexibility in the design, this turned out to be a much easier task than it would have been with a conventional rubble mound structure. This is partly due to the differences in slopes, the front slope versus the core slope, and partly by adjusting the parameters of armour width, A_h, and berm width, B.

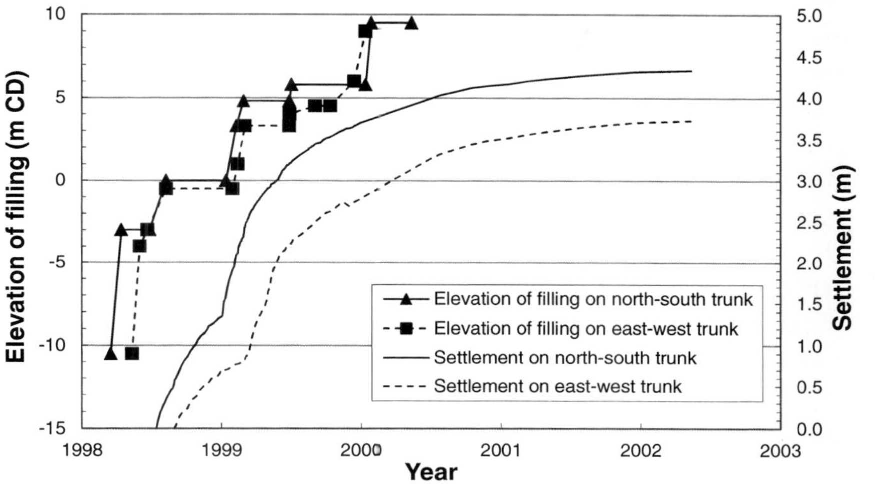

Figure 5.24. Settlement monitoring of the Hafnarfjördur breakwater. The elevation of filling of the two breakwater trunks is shown on the vertical left axis while the settlement is shown on the vertical right axis.

5.6 Maintenance aspects

Maintenance during the life-time of a structure is very common. But for breakwaters one often would like to have a structure that does not need maintenance with respect to armour layer and crest level, as these can be costly to repair. This also applies for berm breakwaters. The guidance in Section 5.2.2 on resiliency serves the purpose of minimising maintenance: the berm should be large enough that even under overload conditions part of the berm remains and repair is not needed after such an event.

Experience with the earlier designed fully reshaping berm breakwaters has shown, however, that the need for maintenance is often under-estimated. The behaviour of the berm was investigated in small or large scale physical wave facilities for design conditions as well as overload conditions and the cumulative effect of these storms were observed. But the effect of many storms below design conditions, may be at slightly different water levels, each contributing a little to the cumulative recession of the berm. Then, after a number of years, the full berm has reshaped and maintenance is needed, by adding rock to the berm. It means that the former designs of fully reshaping berm breakwaters sometimes did not have enough resiliency in the design.

Examples have been described in Chapter 9, specifically Section 9.5 on fully reshaping berm breakwaters. The breakwater at St George in the Bering Sea served quite well for more than twenty years, but had to be supplemented by adding rock onto the berm 21 years after construction. The Bakkafjordur berm breakwater in Iceland had to be repaired after 10 years of service. The Mortevika berm breakwater served for 15 years, but had to be maintained then after a severe storm. The Mackay berm breakwater had to be repaired just after 10 years of service. All of these fully reshaping berm breakwaters had design stability numbers close to or even exceeding a value of 3.0. Information on these breakwaters is given in Section 9.5.

The design guidance on resiliency and acceptable berm recession for fully reshaping berm breakwaters in Section 5.2.2 will lead to a less vulnerable design as for the original designs from the eighties and ninetees of the last century. But it is clear that this type of berm breakwater is less resilient than the partly and hardly reshaping berm breakwater designs. With guidance on getting the largest rock from a quarry in Chapter 6, it is very often possible to design for the more resilient partly and hardly reshaping structures.

Chapter 6

Armourstone and Quarrying

6.1 Introduction

The availability of large armourstone is a very important aspect in the planning and design of berm breakwater projects as well as for all rubble mound structures. This is particularly true for the design of the Icelandic-type berm breakwater, where the information on the availability of large armourstone is regarded as equally important as the information on the wave loads the structure will be exposed to. When using a dedicated quarry for a project, there are often large economic benefits in matching the required volumes for the structure with the expected yield from the armourstone quarry, both with regard to the largest armourstone, as well as the whole yield curve from about 0.5 t or 1 t and upwards.

Some of the guidance on armourstone and quarrying as well as on construction aspects has been published in conference proceedings, like Sigurdarson and Van der Meer [2015]. This chapter summarises most of it, but will elaborate more on most aspects.

Quarry yield prediction has played an important role in the design phase of harbour breakwater projects in Iceland since the early 1980s. The prediction is now mainly based on analysing drilled cores from the potential rock mass, but earlier work was based on only open rock surfaces. Quarry yield prediction has proven to be a valuable part of the design process in preparation for successful breakwater projects.

Without proper quarry yield prediction, the design team often has to rely on contractors or quarry operators for information on the maximum quarry yield or the size of the largest stones obtainable from the quarry. These estimates are very often biased by the size of equipment the contractor or quarry operator has available for sorting and handling of the blasted material.

Dedicated armourstone production is not common and therefore there are not many contractors that have much experience in this field. Guidelines for blasting for armourstones are insufficient and only few contractors have much experience in drilling and blasting for large armourstone. It is therefore important that the design and supervision team has the expertise to supervise the quarry management.

In many countries armourstone quarries are believed to yield only up to 6–8 t armourstone and they rarely yield larger stones. The question is whether this is the practical upper limit of the quarries or if the armourstone market or projects in these countries has not encouraged production of larger armourstone? Therefore the quarry operators have not been challenged to produce larger rock and that might be the reason for the anticipated 6–8 t upper limit.

In Iceland the quarrying for armourstone has developed considerably over the last decades. With improved blasting technique and handling of quarried material, the maximum size of armourstone has increased. In the 1980s designers were not using larger armourstone than 10 t. In the 1990s several projects were performed using armourstone up to 16 t. At the turn of the century the Icelandic team was working on several projects exposed to high waves. Based on thorough quarry investigations, the designers felt confident to gradually increase the maximum size of armourstone, first up to 20 t, then to 25 and 30 t and in the year 2002 up to 35 t. In all cases the projects were tendered out, the contractors were able to quarry the required armourstone sizes and the projects finished without claims on the quarrying.

Recent guidance and literature, [Rock Manual, 2007] and [Coastal Engineering Manual, 2006], has highlighted that variable yield results from armourstone quarrying can be improved by a number of important measures. Recent cases, [Sigurdarson *et al.*, 2005-a and 2005-b], illustrate that unlike blasting associated with aggregates and mining operations, optimisation of the extraction process has to have a focus on the potential for production of large blocks for armourstone right from the outset of the quarry development.

Large armourstones will not be available from the blasting pile unless it is properly planned and the contractor executes blasting and linked production activities appropriately, typically with the technical assistance of the design/supervision team and others with experience in producing large armourstone.

It is also important to realise that breakwater projects can often gain from a small margin of the largest stone class, smaller than 5% of the total mass. It has been experienced several times that although a quarry was generally regarded as having an upper yield limit of 6–8 t, it was possible to get stones of up to 20 t in small margins. With small margins of an armourstone class of say 10–20 t, it is possible to increase the strength of the breakwater considerably compared to upper limits of the largest stone class being 6–8 t.

With regard to the armourstone quarry the relationship between the client and the contractor needs careful consideration. In areas where there is an open market with armourstone and it is not possible to develop a dedicated quarry for the project, it is the contractor's risk to acquire rock for the project. In areas where this is not the case it is often in the client's interest to plan a quarry for the project, make quarry investigations including quarry yield prediction and to include this information in the tender documents. Evidently this involves a risk, both for the client and for the contractor. For the contractor the risk is whether she/he is able to produce armourstone required for the project, which often follows the expected yield of the quarry. The client's risk is mainly associated with the contractor not achieving the quarry yield prediction that was included in the tender documents.

In that case, the contractor not achieving the predicted quarry yield, the question arises, whether the contractor is applying best practice in the quarry or if the yield prediction was too optimistic and misleading. One must keep in mind that afterwards it is not possible to prove that the contractors quarry management was not optimal with regard to armourstone production and that in some cases the contractor could have an interest in not achieving the quarry yield prediction. Very often the demand for the different stone classes is not uniform in time through the whole project. If the project is utilising the full potential of the armourstone quarry this means that heavier stone classes have to be stockpiled during the first phases to be used in the last phases of the project. This is not always in the interest of the contractor.

It is therefore important that the contractor has confidence in the quarry investigations, including the quarry yield prediction, and is willing to adjust the blasting design to fit the required armourstone sizes and volumes. At the same time it is equally important that the client or designer is confident that the contractor is managing the quarry well by using a knowledgeable supervision team.

As a result, and in line with the recognised guidelines [Coastal Engineering Manual, 2006], it is recommended that a range of measures be adopted in addition to standard practice both from the contractor's side as well as from the client's side to ensure that the risk of lower than expected yield is managed. These include:

- use of experienced blast designers with demonstrated and suitable armourstone production techniques;
- use of well-trained inspectors familiar with blasting procedures, stone quality, and stone inspection techniques employed on site;
- use of qualified experts with armourstone production experience, who should monitor and modify blast planning to maintain yield predictions;
- use of qualified personnel with experience in production of large armourstone as a part of the client's supervision team.

Measures such as the above and the use of contractors experienced in large armour production and construction of large rubble mound breakwaters during the procurement and construction phases, will minimise the risks associated with armour size reduction during quarrying and handling.

6.2 Armourstone gradings

6.2.1 *Introduction*

This section discusses armourstone gradings from different viewpoints. Quarries produce rock in various sizes and often up to a certain maximum size. In breakwater design, it is possible to use rock from existing quarries in some cases, which often operate supply-based, where blasting and sorting for armour rock has been optimised to win gradings of rock in specified classes. In other cases it is possible to open and operate a dedicated quarry for the project. The rock gradings for the breakwater project may then be designed based on the predicted yield of the quarry. History shows that specifications for grading curves of rock armour may differ whether the rock comes from existing quarries or from dedicated quarries. Both will be described in this chapter. A third set of specifications for rock grading comes from research in physical scale models.

6.2.2 *Grading curves in research*

Rock gradings used in research in physical scale models are not determined by limitations of quarries as most rock used in research will be less than 1 kg. The gradings used are more likely determined from available rock material in the laboratory or from aggregate sources.

For project-related scale model investigations one should reproduce the grading, on a smaller scale, as specified for the project. But sometimes physical scale models are used for applied research, which is not project-related. Those gradings are described in the present section. Standard gradings according to the Rock Manual [2007] and EN 13383 [2002] for supply-based quarries and gradings for dedicated quarries will be described in Sections 6.2.3 and 6.2.4, respectively.

The governing parameter in stability of rock is the median mass M_{50} and this is defined as 50% *by mass* of armourstones passing. This median mass gives the nominal diameter, D_{n50}, which is used in the stability number $H_s/\Delta D_{n50}$. Another parameter, which is used to describe the grading, is the gradation D_{85}/D_{15} for sieve sizes, D_{n85}/D_{n15} or M_{85}/M_{15}, both on mass. A larger ratio gives a wider gradation. The rock grading in applied research is often given as a "straight line on a log-linear graph", see Figure 6.1 [Van der Meer, 1988-a] and [Thompson and Shuttler, 1975]. The most important area with respect to stability is the curve between 15% and 85%, giving the M_{50}.

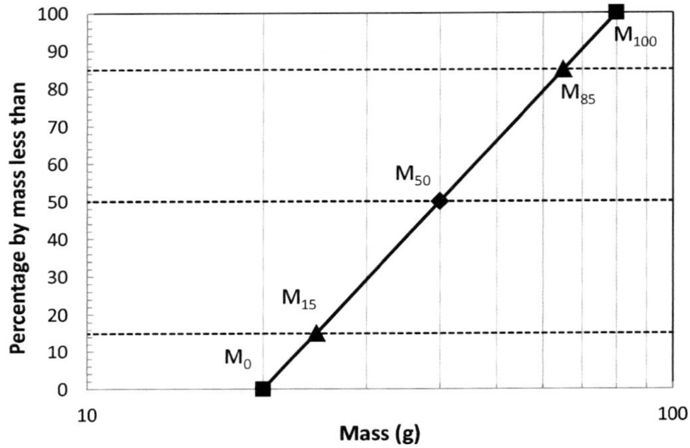

Figure 6.1. A specified grading curve for applied physical scale model investigation, showing a rock class 20-80 g with M_{50} = 40 g (Equation 6.2) and M_{85}/M_{15} = 2.64.

Given the rock class, one can calculate the other masses, like M_{50} or M_{85} and M_{15}. Assume that M_0 and M_{100} give the class limits (it is also possible to assume M_{85} and M_{15} as class limits). Each mass is then given by:

$$M_y = M_0 \exp\{y \ln(M_{100}/M_0)\} \tag{6.1}$$

For M_{50} this means:

$$M_{50} = M_0 \exp\{0.5 \ln(M_{100}/M_0)\} \tag{6.2}$$

If M_{85} and M_{15} are assumed to give the rock class and again a straight line is assumed on a log-linear graph, M_{50} can be calculated by:

$$M_{50} = M_{15} \exp\{0.5 \ln(M_{85}/M_{15})\} \tag{6.3}$$

One can also substitute M_x by D_{nx} if one wants to work with nominal diameters.

Figure 6.2 gives the measured grading curve of a physical model rock sample, where 100 individual stones were weighed. Between 15% and 85% the curve is indeed almost straight, giving $M_{50} = 270$ g and $D_{n85}/D_{n15} = 1.25$. Between 85% and 100% the curve is also fairly straight, except the smallest 10% give some smaller rock.

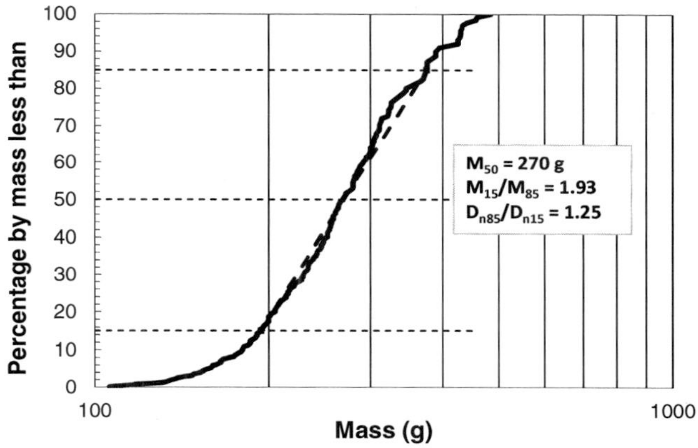

Figure 6.2. Measured grading curve for physical model research, showing a straight line between 15% and 85%.

When comparing test results from various investigations, one should verify that the given definition of M_{50} in this section has been applied to all investigations considered. Still in many quarries there is a potential to produce larger rock and if the standard rock gradings are a limiting factor the design team is encouraged to check upon this.

6.2.3 *EN 13383 system for standard gradings*

The European standard for armourstone, EN 13383 [2002], specifies a system for standardisation of gradings for armourstone, which is also presented in the Rock Manual [2007]. One is referred to Section 3.4.3.2 of the Rock Manual for a description of standard gradings. Of interest here are only the heavy gradings, sizes appropriate for armour layers.

The system is based on setting limit values with an associated percentage passing by mass.

- ELL, Extreme Lower Limit – the mass below which no more than 5% passing by mass is permitted;
- NLL, Nominal Lower Limit – the mass below which no more than 10% passing by mass is permitted;
- NUL, Nominal Upper Limit – the mass below which no less than 70% passing by mass is permitted;
- EUL, Extreme Upper Limit – the mass below which no less than 97% passing by mass is permitted.

It also gives limits for M_{em}, defined as *effective mean mass*, i.e. the average mass by number of a sample of stones without *fragments*, where fragments are defined as stones below ELL.

The EN 13383 standard gradings are defined with relatively wide limits to the grading curves allowing armourstone to be both smaller and larger than the lower and upper nominal limits. Take for example the 10–15 t armourstone class (see Figure 6.3). The standard allows up to 10% of the armourstones to be smaller than 10 t (NLL) and up to 5% to be smaller than 6.5 t (ELL), as well as up to 30% to be larger than 15 t (NUL) and 3% to be larger than 22.5 t (EUL). There are two main reasons for this. One is that it enables the armourstone producer to increase his yield in the armourstone class, and the other is that during handling and transport, some armourstones may break and the wide grading allows the producer to take this into account.

The *effective mean mass,* M_{em}, is not specified precisely, although it is the governing parameter for design of rubble mound breakwaters, including berm breakwaters. The range for M_{em} is given by the Rock Manual [2007] as 12.0–13.0 t for the grading 10–15 t. The relationship between M_{em} and M_{50} for the standard gradings, given in Table 3.6 of the Rock Manual [2007], shows little difference for the heaviest grading 10–15 t, increasing towards the lighter gradings. For the standard gradings M_{50} theoretically lies near 0.5*(NLL+NUL), which would give M_{50}=12.5 t in the example above and in Figure 6.3.

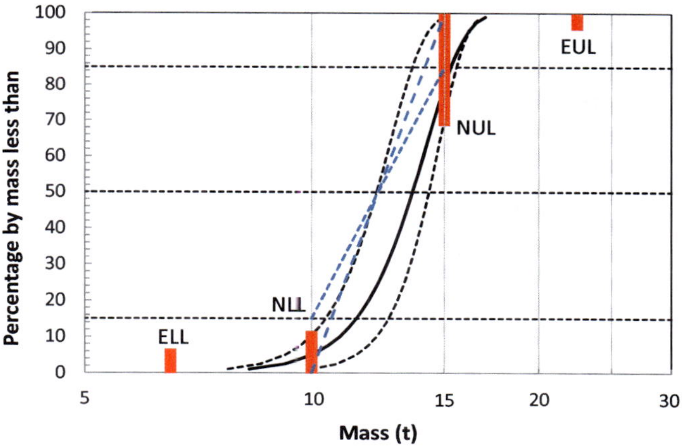

Figure 6.3. Specifications for a rock class of 10-15 t according to EN 13383 [2002] and Rock Manual [2007] compared with gradings based on a straight line as in Figure 6.2 (blue dashed lines).

When the grading curves in Figure 6.3, according to the Rock Manual, are compared with the grading curves method in the previous section (a straight line through M_0 and M_{100} or through M_{15} and M_{85}; see the straight blue dashed lines in Figure 6.3), it appears that the M_{50}-values according to the Rock Manual are equal or larger than found by a straight line. This means that if a design with a stone class of 10–15 t as in Figure 6.3 is tested in a physical model with a gradation curve of same type as in Figure 6.2 and ordering the rock grading according to EN 13383 [2002] this would give a slightly conservative design.

The heaviest EN 13383 standard grading is 10–15 t. In many quarries, there is still a potential of producing larger armourstones and this should be looked at.

There might be a disadvantage in design using the standard gradings. If the required M_{50} from design calculations is slightly above the M_{50} of a specific standard grading, then one will choose to go one step up. Say a M_{50} of 6 t is required from calculations, then one would have to choose a standard grading of 6–10 t. Another option would be to go for a non-standard grading of 4–8 t or 3–9 t. Although it would be on the safe side to stick to the standard gading, it can lead to uneconomical utilisation of the quarry. It may also mean that different safety factors are used, depending on the required M_{50} and the available standard grading.

While the use of standard gradings can in some cases lead to uneconomical utilisation of a quarry in a specific project, there is an advantage, especially for the producer, who can keep a steady production and produce the standard gradings on stockpile. Then, with an order for one or more standard gradings, large quantities can be delivered in a short time.

Another aspect, and maybe a disadvantage, is that the standard gradings become narrower if the rock class becomes heavier. A standard grading of 6–10 t with M_{50} close to 8 t is very narrow with a M_{max}/M_{min}-factor of only 1.7. In comparison, a non-standard grading of 4–12 t also with M_{50} close to 8 t is wider with a M_{max}/M_{min}-factor of 3.0 and has a higher yield from the quarry. For many quarries the difference in yield would be more than double.

The setup of the EN 13383 [2002] standard gradings originates from the rock market around the North Sea and allows breakage during transport, where armourstones are handled several times and transported long distances, often by sea. Often these projects make only use of one or few rock classes. Broken stones not fulfilling the NLL or ELL criteria would then not be used in the project if the grading curve does not allow this.

6.2.4 *Non-standard gradings from dedicated quarries*

Often breakwater projects involve the use of dedicated armourstone quarry for provision of material for the breakwater construction. Then the designer has more opportunities to define non-standard gradings that fit both the quarry capabilities as well as demand for the structure. All size grades from the quarry should be used in the design, often the whole yield curve from the lightest up to the heaviest stones chosen for the project. In these projects the designer does not have to follow the standard EN 13383 gradings and there can be considerable advantages not to do so.

As described above, the standard gradings allow for up to 10% of the material to be lighter than the NLL and up to 5% lighter than the ELL, which weighs roughly about 0.7*NLL. This is reasonable in the market where the standard gradings originate, but it is not necessary and makes projects more difficult to manage when stone classes are used, utilising the whole yield curve.

Stone classes for dedicated quarries should cover the whole yield curve from about 0.3 or 1 t, depending on the project, and up to the largest armourstone chosen for the project. Each armourstone class can be defined according to minimum (M_{min}) and maximum (M_{max}) mass limits and depending on the project either the mean or the median mass, M_{50}. The mean mass is defined as 50% by mass of armourstones passing (Figure 6.1) and the median mass is defined as 50% by number of armourstones passing. To take account of small deviance in sorting, it is practical to allow for 5% of the armourstone to be lighter than M_{min}, but limited to no armourstone being lighter than 90% of M_{min}. These deviances could be due to breaking of stones, inaccuracy in the weighing or calibration of scales or weighing equipment. Note that these limits are more strict than from dedicated quarries and that also the method of mass by number of armourstone passing can be used. Generally, it is permitted for armourstones to be larger than the upper limit, providing that this does not affect the quality of the placement or achieving the filter criteria.

Standard gradings are produced up to the EN 13383 [2002] standard with considerable effort in documented sorting of the material, not only into the stone classes but each stone class has to be sorted into subclasses to ensure that the required median mass is met.

In projects with dedicated quarries, there can be a great economical advantage in transporting the stone classes directly from the blasting pile to the breakwater without stockpiling and placing each class into sub-classes. To facilitate this it is practical to define the required minimum median mass, M_{50min}, as easily achievable within the natural grading from the quarry between M_{min} and M_{max}. as:

$$M_{50\,min} = M_{min} + 0.33*(M_{max} - M_{min}) \quad \text{(by number passing)} \qquad 6.4$$

As the natural grading of armourstone from a quarry within the M_{max} and M_{min} limits results in a higher M_{50} than M_{50min}, no extra measure is necessary to fulfil the M_{50min} requirement. This can be followed up by continuously weighing and recording individual stones. Designing by M_{50} and choosing the stone class by Equation 6.4 results in a slightly conservative design as most quarries yield M_{50} higher than the minimum requirement. The advantage for the contractor (that should reflect the price) is that it is often possible to place armourstones directly from blasting pile onto the breakwater without stockpiling. Where such stockpiling is necessary, the costly management of sub-classes can be avoided.

The minimum grading curve for each stone class is then defined by the following points, where M_{50} is based on number passing:

- $M_0 = 0.9*M_{min}$
- $M_5 = M_{min}$
- M_{50} = Equation 6.4
- $M_{100} = M_{max}$

It is becoming common in large projects that the contractors choose to execute the sorting of armourstone from the blasting pile with equipment (front loaders or excavators), equipped with loadcells to weigh and register the armourstone. The contractor keeps daily records of M_{50} for each stone class. Usually no extra measures are required to fulfil the M_{50} requirement than just keeping all stones within the M_{min} and M_{max} for each class. If this is not the case, it is easily noticed and the measures could then be to take out some of the smaller rock. In the Hammerfest project, for instance [Sigurdarson *et al.*, 2005-a], as a part of quality control, the contractor chose to weigh and record all armourstone larger than 4 t.

6.3 Quarry yield prediction

For the designer of a rubble mound structure, it is necessary to know what sizes of armourstone can be used in the design. In a moderate wave climate, where armourstone can be sourced from operating quarries producing the standard gradings, this is no problem. But things get more complicated when either the wave climate requires armourstone larger than the standard gradings or when there are no operating quarries nearby. Then it becomes necessary to predict a workable quarry yield for the project, either of armourstone larger than usually produced from the operating quarry or from a rock mass available from a dedicated armourstone quarry.

The first step in predicting a workable quarry yield is to assess the *in-situ* block size distribution. That is the distribution of the natural block sizes in the rock mass prior to quarrying. The *in-situ* block size distribution is determined by the spacing between discontinuities cutting through the rock mass. These can be natural joints, bedding planes, other natural fracturing and weakness planes. Figure 6.4 shows the fracture pattern of the rock used for the Sirevåg breakwater.

Figure 6.4. Coarse fracture pattern of the anorthosite gabbro rock used for the Sirevåg berm breakwater.

In armourstone quarrying the aim of the extraction process, which is usually blasting, is to loosen the rock mass by opening up the natural discontinuities and to produce a workable blast pile. Inevitably the fracturing by the energy release from the explosives not only opens the natural discontinuities, but also opens new fractures. Some of the *in-situ* natural blocks will be divided into smaller blocks. But while the *in-situ* block size distribution is uncontrollable, the degree of fragmentation during blasting is controllable.

Several methods have been presented to predict the *in-situ* and the blasted block size distributions. A summary of them is given in the Rock Manual [2007] and is not repeated here. Common to these methods is that they are mostly based on the mean spacing between discontinuities, joints and fractures, and consequently, they result in the average block volume or average block weight of the whole quarry.

The designer is not interested in the average block weight, he is interested in the block sizes from the heavier end of the grading curve and often the small fraction of largest possible armourstone from the rock mass. In the Hammerfest project, with a design wave height of H_s=7.5 m, the largest stone class used in the design was 20-35 t and yielded 3% to 5%, see Sigurdarson *et al.* (2005-a and 2005-b). But the average yield from the quarry, 50% by mass or volume, was only about 0.1 to 0.2 t. Obviously, information on the average yield from the quarry would not help the designer very much. Figure 6.5 shows the stockpile of the two largest armourstone classes for the Sirevåg breakwater.

Figure 6.5. Stockpile of Class I, 20–30 t, and Class II, 10–20 t, armourstone for the Sirevåg breakwater.

The quarry yield prediction developed in Iceland is initially focussed on the heavier end of the grading curve, but has also been adapted to deal with the lighter end of the graph as well. Omar Bjarki Smarason was the originator of the method [Smarason et al., 2000]. His yield predictions aided his Icelandic colleagues in their development of the Icelandic type berm breakwater and examples have been described further on. Through experience in a number of armourstone projects over a period of more than 30 years, the quarry yield prediction has been proven useful for igneous rocks of all types, ranging from fractured basalt lavas to more solid gabbro and granite intrusions. It has also been applied in projects worldwide in various quarries consisting of rock of sedimentary origin as well as flow banded metamorphic rocks.

Figure 6.6. Drilled borehole cores which have been analysed for quarry yield prediction of a breakwater project in Husavik, Iceland.

The method relies on logging of discontinuity spacing in drilled cores. The discontinuity spacing is used to calculate the rock quality designation value, for various space intervals up to 2-3 m, in order to get an indication of the possible bench yield of the quarry. Together with the fracture frequency it is used to work out a primary blasting curve and consequently a secondary yield curve to fit the desired maximum size of the design. Originally Smarason's analysis relied on scan-lines in open surfaces only, as drilled cores were not available for most small projects until the early 1990s. Since then the method relies on drilled cores only, see Figure 6.6. The rock quality designation value is defined as the proportion of the scan-lines that consists of intact lengths of 0.1 m or longer. The quarry yield prediction uses not only rock quality designation values based on 0.1 m, but also on other core lengths up to 2 m or even longer and represents an in-situ block size prediction based on this information.

When logging the core, it is important to distinguish between discontinuities that will open during the fragmentation process and handling of the stones from the blasting pile to the breakwater, and those discontinuities that will not open. This is especially important when predicting maximum workable armourstone sizes from the rock mass, which will only be done with manual logging of the cores. Although the logging of cores is usually based on vertical cores, it may sometimes be supplemented with inclined cores, depending on the rock types and nature of discontinuities and bedding layers.

The quarry yield prediction is then determined by shifting the *in-situ* block size distribution, to account for various effects. The volume reduction or degree of shifting depends on various factors of the blast design as well as several site and rock conditions. It is also different for different parts of the yield curve, the light and heavy parts of the grading curve and includes a compensation for further splitting due to handling of armourstone from the blasting pile to the breakwater.

The quarry yield prediction for a project in Hornafjörður, Iceland (Figure 6.7), is an example of a quarry yield prediction derived from *in-situ* block size distribution, [Smarason *et al.*, 2000]. This distribution is given by the black line in the graph and is based on the above mentioned logging of spacings of drilled cores. Two quarry yield predictions are then presented, A and B, where prediction B takes only the better parts of the quarry, while prediction A should be representative for a larger area, including poorer parts. These are the yellow lines in the graph.

According to these predictions the yield of armourstone heavier than 10 t would be 10 to 15% of the total quarried volume, while it is 21% in the unblasted rock, according to the joint space average (*in-situ* block size curve in the graph). With a design wave height of H_s = 3.8 m an

Icelandic-type berm breakwater could have been designed with a maximum stone size of 5 t. But as larger stones were available from the quarry these should be used to increase the stability of the structure. It was decided that the two heaviest stone classes should be Class I 8–15 t and Class II 5–10 t. Figure 6.7 also shows a design curve, given by "Required design volumes" and a curve called "Produced from quarry", which is the achieved yield from the quarrying. Note that only 4% in volume was needed for stones larger than 8 t. This is the reason why the curves of required and produced volumes are much lower in this area than the yield predictions.

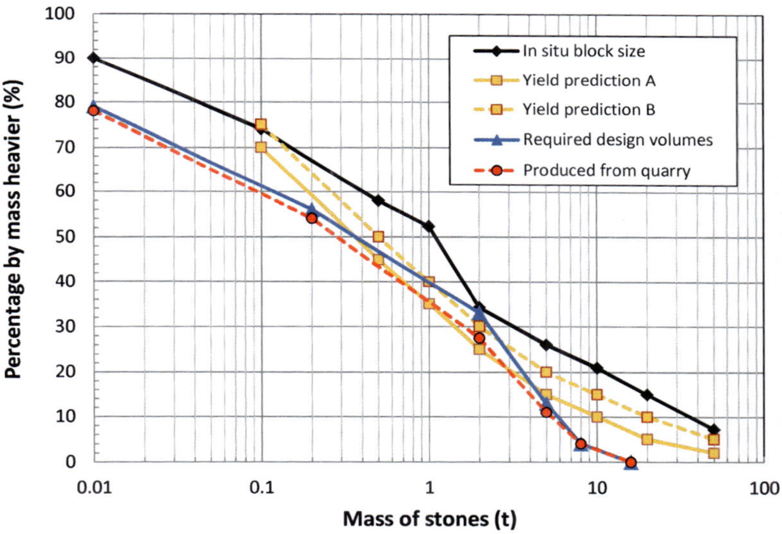

Figure 6.7. *In-situ* block size distribution, quarry yield predictions, required volumes and production results for the Halsendi quarry for the East Barrier project, Hornafjördur, Iceland. Re-drawn from Smarason *et al.*, [2000].

Predicting a quarry yield from the *in-situ* block size distribution of a rock mass can be done with various degrees of achievability. Easily achievable quarry yield predictions assume that not only will all discontinuities open during the fragmentation process, but in excess to that most of the larger natural blocks will break up into several smaller blocks. But with knowledgeable blasting techniques and handling of armourstone, it is often possible to achieve a much higher yield of large armourstone. Challenging the quarry management and operation is to make a quarry yield prediction close to the upper limit of achievable quarry yield. This requires a yield prediction based on a careful analysis of the discontinuities and an understanding of the process.

Figure 6.8 shows another example of quarry yield prediction, in this case for the Laukvik breakwater in Norway, [Sigurdarson *et al.*, 2006]. According to the average *in-situ* block size distribution, 33% of the unblasted rock by volume or mass is larger than 10 t. After blasting, the predicted maximum yield above 10 t became 22%. For the more conservative "Adjusted yield prediction" the predicted yield above 10 t became 17% with an upper limit to the stone size of 50 t. The required stone volumes are given by the blue lines and are well below the predicted yield curves. The maximum stone size required was 30 t.

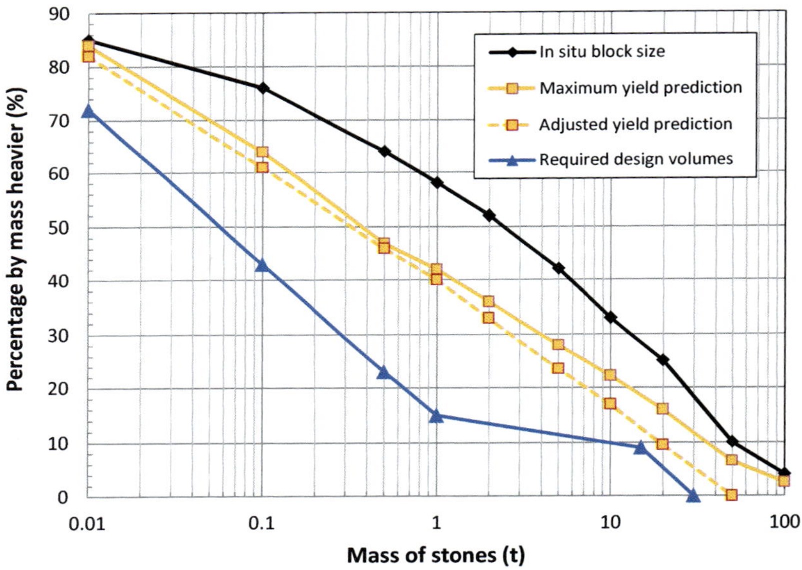

Figure 6.8. *In-situ* block size distribution, quarry yield predictions and required volumes for a proposed quarry for the Laukvik breakwater. Re-drawn from Sigurdarson *et al.*, [2006].

Test blasting or trial blasts to determine the quarry yield are not recommended. The quarry yield prediction, based on analysis of drilled cores from the rock mass with a reasonable grid, is far more reliable than test blasting. Test blasting is limited to a small part of the rock mass to be quarried and there can be considerable variability in the rock mass. Often it is necessary to go deep into the rock mass for the larger armourstone.

Therefore, a test blasting near the surface of the quarry may give misleading results. As will be explained in the next section, the optimum blasting programme needs to be tuned to the special rock mass and it is unlikely that this can be done during the trial blasting. It is also necessary to bring in large equipment to sort material from the blast pile. It is unlikely that this will be done for a test blast. Test blasting in an open quarry can give valid information if full size equipment is brought in for working the blasting pile. On the other hand, opening a test quarry in greenfield sites is not recommended.

6.4 Blasting for armourstone

Normal blasting practice for rock and for aggregates differs from blasting for armourstone. The main principle in blasting for armourstone is to blast with low specific charge with the aim of opening up existing fractures and joints and creating as few new fractures as possible. At the same time the aim is to produce a blast pile that is easily workable by the excavator or front loader.

Figure 6.9. Blasting pile in the harbour excavation area of the Hambantota Port Development project, Sri Lanka.

In the Hammerfest project the best results were obtained by blasting with minimum power often with a large part of the blast pile still standing, which would be considered a failure in conventional blasting. Figure 6.9 shows the blasting pile in the harbour excavation area in Hambantota, Sri Lanka, after a test blast, where most of the blasting design parameters had been changed to improve the yield of 5–10 t armourstone, [Sigurdarson *et al.*, 2014]. One row was blasted instead of multi rows, the specific charge had been lowered, the space-burden ratio decreased, and the distribution of charge within the blasting hole was changed. With a bench height of about 12 m the blast pile stands about 2 to 3 m lower after blasting and is easily workable. The brownish colour of the stones shows that the blasted rock is mainly split by the natural joints and fractures in the rock mass.

A good guideline for blasting for armourstone is the Rock Manual [2007] and these guidelines have not been repeated here. Other references are the first Rock Manual [1991], Smarason *et al.* [2000] and Sigurdarson *et al.* [2005-a]. With increased experience from a number of projects and production of large armourstone, the blasting methods have developed.

The main considerations in blasting for large armourstone and recommended range of blasting parameters are described in the following list, where the Rock Manual [2007] gives definitions. Some definitions of bench blasting terms are given in Figure 6.10.

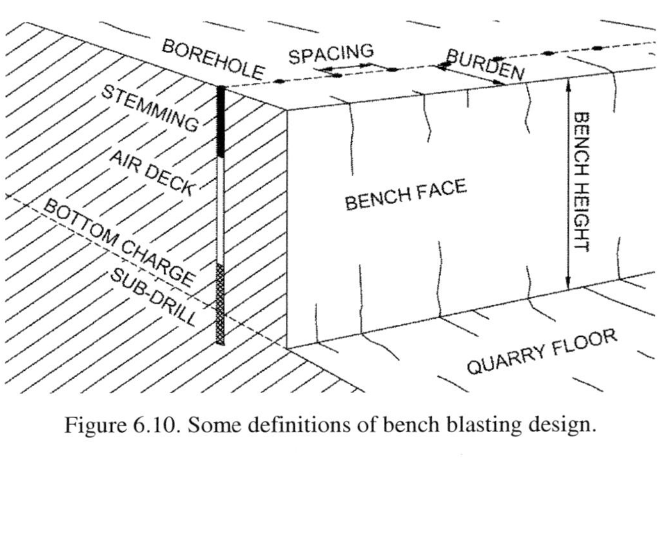

Figure 6.10. Some definitions of bench blasting design.

- A low specific charge, generally less than 200 g per cubic metre of solid rock. Often the best results have been achieved for specific charges close to or even lower than 120 to 150 g/m^3;
- The low shock energy explosive ANFO (ammonium nitrate-fuel oil) in combination with dynamite and slurry (water gel explosive) is usually recommended;
- The presence of water in the boreholes affects the choice of type of explosive;
- The explosives should be arranged as bottom charge. Column charge is not recommended;
- A large stemming length, possibly in a combination with an air-deck, is recommended;
- Usually the stemming length is larger than 0.7 times the burden, also seen in Rock Manual [2007], Figure 3.55;
- Usually a sub-drill of around 0.5 m to 1 m is recommended;
- The usual diameter of drilling bits is between 63 mm (2.5") and 100 mm (4"), depending on the size of armourstone and bench height;
- The optimum spacing-to-burden ratio is less than 0.8, preferably close to 0.7;
- The optimum borehole area lies between 8 m^2 to 20 m^2, depending on the borehole diameter, the bench height, the required armourstone sizes and the discontinuity spacings;
- Often, it is chosen to tilt the boreholes by a few degrees;
- It is critical to maintain high drilling accuracy in blasting for armourstone. Inaccuracy in borehole alignment might lead to a toe formation (unbroken rock on the quarry floor);
- If not constrained by other factors, the optimal bench height is between 8 m and 15 m, depending on the burden, the size of equipment working on the blast pile as well as the required armourstone sizes;
- Drill and blast only one row at the time. Blasting more than one row at the time has proven to affect the yield of armourstone considerably;
- Clearing the bench face and quarry floor before drilling a new borehole is recommended;
- Blast as many holes as possible instantaneously. Inter-hole delaying will cause more fracturing of the rock mass. If there are constraints on

ground vibration, as from nearby structures, consider splitting the quarry face into segments;

- Avoiding drilling parallel to the major joints of the rock mass;
- The drilling pattern for large armourstone is coarser than for smaller armourstone;
- Secondary breakage of oversized stones may be necessary.

Many of the blasting parameters described above are inter-related and inter-dependent. Therefore the recommendations above should not be used in isolation or without due consideration of the inter-dependency, as well as of various site and rock conditions not mentioned.

Production of armourstone by blasting of bedrock is by no means an easy task and requires great care and discipline. A slight alteration of burden, spacing and tilting of boreholes or the use of explosives, may help to improve armourstone yields. The blasting engineer must therefore be prepared to adjust the blasting design to suit each particular quarry and the target yield of armourstone on a blast-by-blast basis. The blasting design is also adjusted as the quarrying advances. Often a few shots are needed for the main adjustments, followed by fine-tuning through the quarrying process, based on blast assessment and the target yield of the structure.

During production, the performance of the blasting and handling operations should be assessed and compared to the quarry yield prediction and target yield. To support refinement of blast designs, fragmentation of each blast should be measured and recorded by weighing of stones, preferably all of the larger stones. This does not have to be a costly operation if the equipment used to sort material from the blast pile is equipped with scales.

The blasting design in the Hammerfest project mentioned above (see Figure 6.11), was based on experience from other similar projects and was further developed in cooperation with the explosives producer DYNO, [Sigurdarson *et al.*, 2005-a]. The target yield of the heaviest stone classes was 2.5% into Class I (20–35 t) and 3.8% into Class II (10–20 t). The contractor's goal was to blast the rock to get fragmentation as close to the design curve as possible. This was achieved by carefully monitoring: (i) geology; (ii) counting and weighing all stones over 4 t; (iii) drill pattern and bore hole diameter; (iv) type and quantity of explosives.

The best results were obtained: (i) with a hole spacing (E) of 3.5 m and burden (V) of 4.5 m; (ii) with an optimal bench height of 12–14 m; (iii) by blasting only one row at the time; (iv) by avoiding drilling parallel to the major joints; (v) by keeping the drill hole area 8 m^2 to 20 m^2; (vi) by keeping a high charge density at the bottom of holes to secure a good working platform.

The average charge was 170–200 g/m^3 of anfo, dynamite and slurry and the Nonel system was used for detonation. Fragmentation of each blast was recorded by weighing all stones over 4 t. Face mapping was used to help decide the drill pattern and use of explosives. The holes were typically only loaded in the bottom and the middle section was left empty. Sand was used for stemming. This method is referred to as the air-deck method.

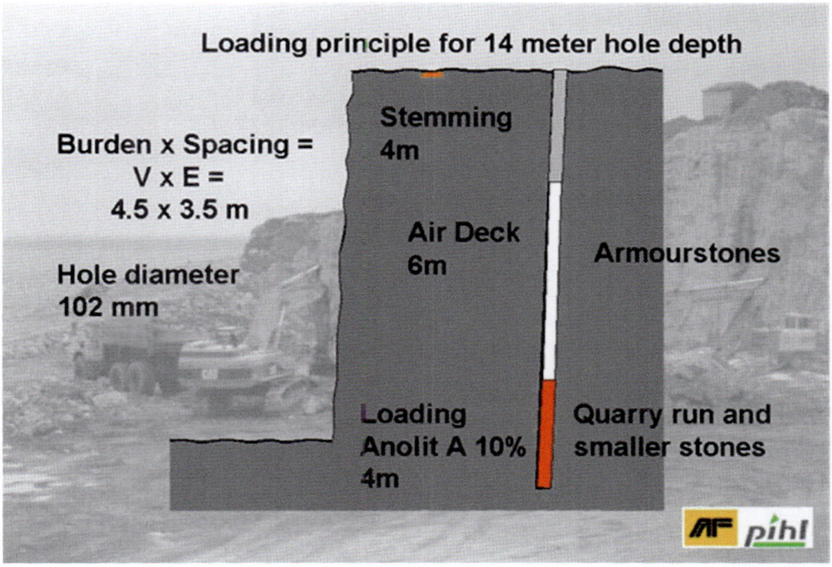

Figure 6.11. The blasting design in the Hammerfest project [Sigurdarson *et al.*].

6.5 Quarry planning

Armourstone quarries have to be planned on basis of the rock source to be exploited. Some rock sources are homogeneous, while others are inhomogeneous, both in vertical and in horizontal directions. A thorough quarry investigation identifying the best suitable rock source is therefore necessary. Depending on the volume of material needed for the construction of the breakwater and the time available for quarrying, a production capacity can be chosen to balance the quarry output to the construction requirements. With multiple operating locations, it is possible to achieve a blast and output cycle to meet the required production capacity. This can be done with multiple faces, a single long face or multiple benches.

Often, pairs of excavators and wheel loaders are used to work on the blasting pile for sorting armourstone in different stone classes from the quarry run and to load the material on trucks for transport as seen in Figure 6.12. In smaller projects, some contractors choose to use only excavators to work on the blasting pile and load on trucks, instead of excavators and wheel loaders.

Figure 6.12. Sorting of material from the blasting pile with excavators and front loaders, at the same time loading of trucks from the stock pile. The Hammerfest project in northern Norway.

Quarry management and breakwater construction has to be planned together. All handling of material, like loading onto trucks, hauling or sorting of material, costs money. It is important therefore to minimise stockpiling, or sorting operations into sub-classes.

In general, all quarry run used as core material should be hauled directly to the breakwater and put in place. With proper planning of the breakwater construction, it is often also possible to place a part of the armourstone classes directly onto the breakwater, especially the lighter classes. Usually the heavy armourstone classes must be stockpiled, as these will be used in the final stages of construction.

To be able to sort material directly to an armourstone class and not into sub-classes depends very much on the definition of the armourstone class, mainly on the criterion of the median weight, M_{50}. With a relaxed median weight criterion, as proposed in Section 6.2.4, it is possible to skip additional sorting operations and enable direct placement of armourstone from the blasting pile to the breakwater. If the median weight is defined close to the natural gradation between M_{max} and M_{min}, it is often necessary to stockpile all armourstone classes, often in sub-classes, and to adopt additional sorting operations to fulfil the median weight criterion.

If the required size distribution of armourstone for a breakwater project is close to the achievable distribution from the quarry, it is necessary to plan the armourstone production with regard to the large armourstone from the start of blasting. Usually the largest armourstone cannot be placed until close to the end of construction and therefore need to be stockpiled. These are often not more than few percentages of the total breakwater volume.

If, on the other hand, the quarry yields much better than required for the breakwater, the contractor can choose to produce the large armourstone late in the construction period. However, this involves the risk that if he is not able to produce to the expected quarry yield, at the end of the project he might have to blast only for the largest armourstone, and if the yield for those is 5% or 10% then he is not using 95% or 90% of the last blasts for the breakwater.

Planning of the quarrying and construction of the Sirevåg berm breakwater in Norway was a challenging task. A new quarry was to be opened in a hilly area with very limited area for stockpiling of material. The area was at a small bay opposite to the breakwater construction side. All material had to be transported by sea and the contractor chose to use a split barge for the task. Figure 6.13 shows the initial construction phases of the Sirevåg berm breakwater. Instead of starting to dump quarry run along the centre line of the cross-section, the contractor chose to start dumping material at the toe. By doing this he was gradually able to dump most stone classes directly to the breakwater instead of having to stock pile them. In the first phase it was soon possible to dump Class IV 1–4 t, on top of Class V, the quarry run. Then in the second phase Class III 4–10 t followed and at last in the third phase Class II 10–20 t. It was only necessary to stockpile Class I 20–30 t.

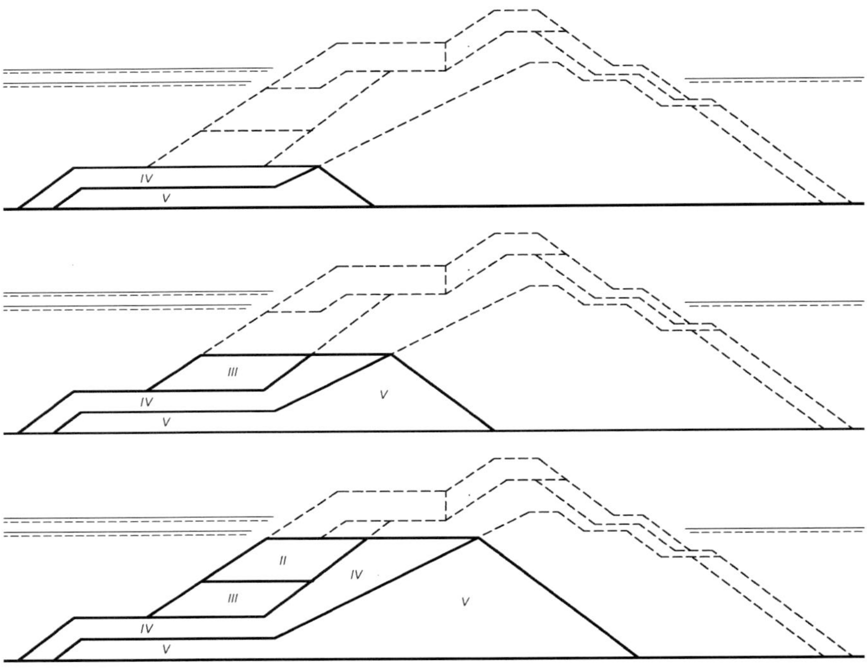

Figure 6.13. The initial construction phases of the Sirevåg breakwater for material dumped by a split barge. Class V is quarry run, Class IV is 1–4 t, Class III is 4–10 t and Class II is 10–20 t.

Chapter 7

Construction

7.1 Introduction

Good manuals such as the Rock Manual [2007] and Coastal Engineering Manual [2006], give general advice on construction of all kinds of breakwaters. This chapter will not repeat the general advice, so the reader is referred to those manuals for general issues on construction. This chapter is dedicated to construction issues applying to berm breakwaters, especially where they differ from conventional breakwaters. These issues are often associated with the use of a dedicated quarry and the design fully utilising the quarry yield of armourstone.

Working in a marine environment often involves higher risk than working on land. Often the client stretches the construction period over to periods of the year where high waves can be expected, perhaps approaching design wave conditions. Evidently this involves a risk that the partly constructed breakwater may suffer damages, but experience suggests that partially constructed berm breakwaters are "tough" and less vulnerable than conventional breakwaters. This is partly due to wider tolerances under water and partly due to less vulnerable construction methods.

7.2 Armourstone quarry in the contract

When a breakwater project is planned in an area where a market with armourstone is not active, then it is in the client's interest to locate a possible quarry or quarries for armourstone and core material. This can either be an already open quarry or opening a new quarry. In that case the process involves an agreement with landowners, acquiring environmental approval and planning permission for the quarry. This may be a lengthy process and in some cases it is better not to leave this to the contractor to arrange.

A part of the quarry investigations is a quarry yield prediction, discussed in Chapter 6. It is in the client's interest to provide the contractor with as much information as possible about the quarry and the expected yield of armourstone. This involves a risk that the contractor is not able to fulfil the expected yield, either because he lacks the knowledge to work the quarry with regard to armourstone production or because he lacks the will to do so at the particular time of the project. It is therefore important that the quarry yield prediction has suitable safety margins. It is also necessary that the supervision team is familiar with armourstone production and, if the achieved yield lies below the predicted yield, then the team must be able to point out early how the contractor can improve the yield.

If on the other hand, it is proven that the quarry is not yielding as expected, the geometrical design guidelines presented in this book provide tools to act and change the design if needed.

When quarrying for armourstone, transport of material from the quarry to the breakwater and construction of the breakwater, are all the responsibility of the contractor, then it involves less risk for the client to base the contract on volume instead of weight. The volume is less disputable than the weight of material needed to fill that volume. When contracts are in weight, factors such as packing density and layer porosity have a large influence on the bill. But these factors can be difficult to determine. The volume on the other hand, can easily be calculated from cross-sections, but does require a definition of the constructed surface of the rock structure. A contract in weight is more in favour of the transporter or the supplier of rock to the breakwater as it reduces their risk but leaves the client with higher cost risk.

7.3 Equipment

Nowadays, equipment can be rented for a specific project or sometimes bought with a guaranteed selling price at the end of the project. This means that the contractor does not have to own the equipment needed for the project when giving her/his tender, but can acquire or lease the best equipment for the specific project. The project will benefit from a thorough study of construction procedures to determine the most suitable construction equipment.

As for other rubble mound breakwaters, berm breakwaters are constructed both with land-based as well as with waterborne construction equipment. On land, trucks bring material on to the breakwater, quarry run or armourstone. For short distances, front loaders can transport armourstone and dump at locations where long reach is not required. Placing of different layers of armourstone with land-based construction equipment is usually done with excavators, but sometimes cranes are used, especially if there is need for long reach. Compared to cranes, the placement rate of excavators is higher, but other factors such as reach and vulnerability for waves, matters when choosing the best equipment for a specific project, [Sigurdarson *et al.*, 1999].

In some of the earlier berm breakwater projects, like Codroy, Canada, and Hofsos, Iceland, bulldozers were used to push rock onto the berm. This resulted in too many fines mixing with the armourstone class, which diminishes the permeability and is not recommended.

Excavators that have been used for placing of armourstone may range in size from about 20 t to about 120 t. The size is chosen depending on the size or weight of armourstone to be placed as well as the required reach. To allow for heavier lifting capacity to place armourstone in long reaches, the excavator is altered by the dealer or factory by increasing working pressure of the hydraulics and increasing its counterweight. It is assumed that the excavator uses its tipping capacity to its full extent. As a rule of thumb, the excavators can place armourstone to the full reach of a normal boom with a weight of up to one-third of the total weight of the excavator. For example, a 110 t Liebherr R984 excavator with a 7.8 m long gooseneck boom and a 4.5 m long stick (see Figure 7.1), has been used to place a 35 t stone under water at 12 m from its centre point, utilising the buoyancy of the submerged stone. If longer reach is required by land-based equipment, this can either be done with a crane or with an excavator driving on temporary placed stones, see Figure 7.2.

Figure 7.1. A 110 t Liebherr R984 excavator with a rock prong at the Sirevåg breakwater construction.

Figure 7.2. Excavator with a long boom and placed on a temporary level to place rock up to 2 t to a level of -8 m SWL.

The most common waterborne construction equipment is split barges (see Figure 7.3 to Figure 7.5), but side dumpers or side tippers can also be used. Split barges are usually self-propelled. The size varies from about 300 m^3 to 800 m^3, 40 to 60 m long. Split barges have not only been used to place quarry run and smaller stone classes, but also larger armourstones up to about 20 t. In that case only one row of large stones is placed in the center of the barge (see Figure 7.5). Split barges are less sensitive to waves than side dumpers, both during loading, sailing and dumping. With limited space on the breakwater, the land-based operations can be congested. Therefore, using waterborne equipment to place material and moving the loading operation from the breakwater itself can be useful.

Figure 7.3. Loading a split barge with quarry run by wheel loaders. Sirevåg breakwater.

Waterborne placement of material to the lower parts of the breakwater makes it possible to use lower size grades material, such as quarry run or smaller stone classes, in areas that otherwise would be filled with larger size grades, if dumped with land-based equipment. Another advantage of placement with waterborne equipment was described in Section 6.5 and Figure 6.13, where the construction of the Sirevåg breakwater was planned with the aim of being able to place as many stone classes as early as possible instead of having to stockpile them.

In case (part of) the rock has to come from quite some distance from the breakwater site and the quarry is located at the sea side, it is also possible to use large barges to bring the rock to the construction site (see

Figure 7.6). Trucks and excavators are then used to dump and place the core material. An advantage of bringing rock in by sea is that truck transport around the breakwater site is reduced, which is favourable if the harbour is close to developed area.

Figure 7.4. Loading a split barge with Class IV rock by a wheel loader. Sirevåg breakwater.

Figure 7.5. Loading a split barge with large rock in one row by an excavator. Sirevåg breakwater.

Figure 7.6. Large barges (the black boxes on the picture) importing core material, moored on the lee side of the breakwater.

Mass-armoured berm breakwaters can be constructed by dumping of armourstone, usually without any placement requirements. The Icelandic-type berm breakwater, on the other hand, requires the outermost armourstone on the front slope and on top of the berm to be interlocked.

But there is a fine balance in achieving interlocking and not diminishing the permeability. The placement of the armourstone should not be too compact or tight and closing the surface of the breakwater. Often the best equipment to achieve this is an excavator. In contrast to cranes, excavators can turn, rotate, push or pull armourstone to achieve good fitting and interlocking.

Table 7.1. Recently constructed Icelandic-type berm breakwaters.

Project / Location	Construction year	H_s	Class I armour stone	Lowest level of Class I *(m CD)*	Total volume	Deepest section *(-m CD)*
Sirevåg (Norway)	2000–2001	7.0 m	20–30 t	-1.0 m	620,000 m^3	-18 m
Húsavík (Iceland)	2001–2002	6.8 m	16–30 t	-1.0 m	270,000 m^3	-12 m
Grindavík (Iceland)	2001–2002	5.1 m	15–30 t	-4.5 m	170,000 m^3	-5 m
Hammerfest (Norway)	2002–2003	7.5 m	20–35 t	-7.0 m	3,000,000 m^3	-35 m
Vopnafjörður (Iceland)	2003–2004	5.0 m	8–28 t	-2.0 m	140,000 m^3	-9 m
Thorlákshöfn (Iceland)	2004–2005	5.5 m	8–25 t	-4.5 m	230,000 m^3	-5 m
Landeyjahöfn (Iceland)	2008–2010	6.1 m	12–30 t	-2.5 m	600,000 m^3	-9 m
Helguvík (Iceland)	2008–2010	5.0 m	15–25 t	-2.0 m	350,000 m^3	-28 m

The size of excavators varies with weight of the armourstone to be handled. Projects where the maximum stone size is of the order 10 to

15 t, need excavators that weight up to 40 to 50 t. In the last 15 years, several berm breakwaters of the Icelandic-type have been constructed using extra-large armourstone, with maximum stone size exceeding 20 t.

Table 7.1 provides information from these projects in relation to the construction period, design wave height for the most exposed section of the breakwater, largest armourstone class used, total volume and water depth at the deepest section of the breakwater. The largest excavators used to place the heaviest armourstone were about 110 to 120 t for projects at Sirevåg, Húsavík and Hammerfest in the list and about 80 t to 90 t for the other projects.

7.4 Placement and tolerances

7.4.1 *Armourstone placement*

Berm breakwaters are constructed of both "bulk-placed" armourstone and core and for the Icelandic-type also "placed primary armour".

The term "bulk-placed" is defined as armourstone or core material placed in bulk and as for the armourstone without any control of orientation. Bulk-placed armourstone and core shall comply with the following requirements:

- Delivered and placed according to the dimensions, lines, levels and slopes shown on the drawings and within the tolerances detailed in Table 7.2 in the next section;
- Placed to achieve an even distribution of stone sizes without concentrations of smaller stones.

Bulk-placement can be done in many ways, by end-tipping with trucks, with wheel loaders and excavators, with cranes by using rock trays or skips, as well as with floating barges. Each method has its advantages and disadvantages with regard to reach, placement rate, vulnerability to wave and wind conditions, as well as cost. End-tipping with trucks and wheel loaders have shortest reach, cranes have the longest reach, while barges have unlimited reach but are restricted by water depth.

The term "placed primary armour" is defined as individually placed armourstone located within the first two layers on the exposed side of the breakwater, berm front slope, top of berm and on the crest, above an elevation just below the low water level and upwards. In some cases the berm front slope is reinforced with three layers of placed primary armour instead of two. This might be the case for partly reshaping berm breakwaters where the reshaping is expected to come close to two armourstone sizes.

Placing of primary armourstone to increase stability of the berm may comply with the following requirements, which were specified for a specific project of an Icelandic-type hardly reshaping berm breakwater, but which may also be specified for other projects:

- Primary armourstone shall be delivered and placed according to the dimensions, lines, levels and slopes shown on the drawings and within the tolerances detailed in Table 7.3 in the next section;
- Placing shall commence at the lowest level of the section and proceed upwards towards the crest;
- Primary armourstone smaller than the equivalent grading class limit shall not be used to fill interstices, or to prop larger armourstone in order to achieve the required profile. The placement shall also cause minimum disturbance or dislodgement of existing armourstone layers;
- The volumetric porosity for primary armourstone shall be between 0.32 and 0.4;
- Primary armourstone shall be individually placed and a minimum of three points of contact shall be achieved at all times. Primary armourstone shall be placed to achieve a fully interlocked armoured slope, such that each armourstone is securely held in place by its neighbours and placed in such a way that stability is achieved from interlocking and frictional resistance, and not from friction on one plane alone. Interlocking means that each stone is resting firmly on the stones beneath and is in firm contact with all stones surrounding it;

- Special attention shall be made to placement at the intersection of the berm and slope to ensure that a high degree of interlocking is provided. The intersection shall present a curved profile so that there are no abrupt changes in profile or protrusions of armourstone;
- Special attention shall be made to placement at any changes in breakwater alignment (for example the breakwater elbow) and on the roundhead to ensure a high degree of interlocking is provided;
- Tipping of primary armourstone from vehicles, or bulldozing or dumping from hoppers or barges into final position is not permitted. Primary armourstone shall be placed with equipment capable of placing the stone at its final position before release. In addition, all primary armourstone shall be placed with equipment capable of moving and repositioning a released stone if necessary;
- Primary armourstone shall be placed to achieve an even distribution of armourstone sizes without concentrations of smaller armourstone;
- Elongated armourstone, where l/h > 2.5, shall be placed with the long axis normal to the slope;
- In addition, it is recommended that the level of the armourstone is built up progressively and follows a placement sequence similar to that detailed in Figure 7.7

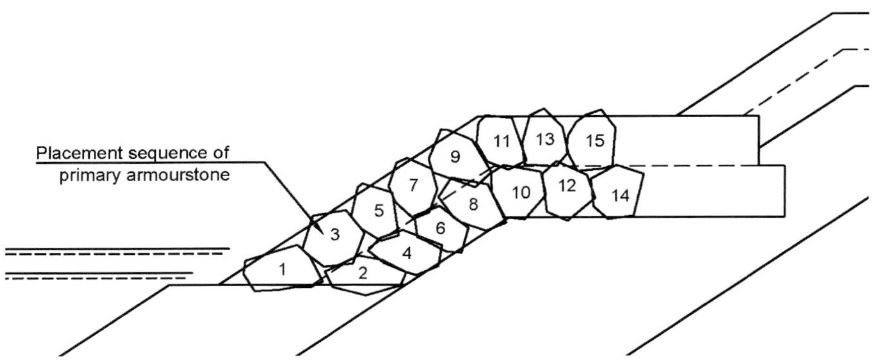

Figure 7.7. Recommended placement sequence of primary armourstone.

The specific placement sequence of primary armourstone, as in Figure 7.7, works best if the underlying rock, be it Class II or even Class I, does not reshape too much and continues to function as a support for the individually placed primary armour. Specific placement of large stones can generally only be done above water. Wave action may be disruptive below this level. The corner of the Class I rock may be reshaped a little, but not enough to loose support for the upper stones.

This example of a strengthened support berm with special placement of primary armour rock worked very well in a physical model test, described in Section 3.6, Figure 3.7, described by Project 1, Test 3 and also given in Table 3.2.

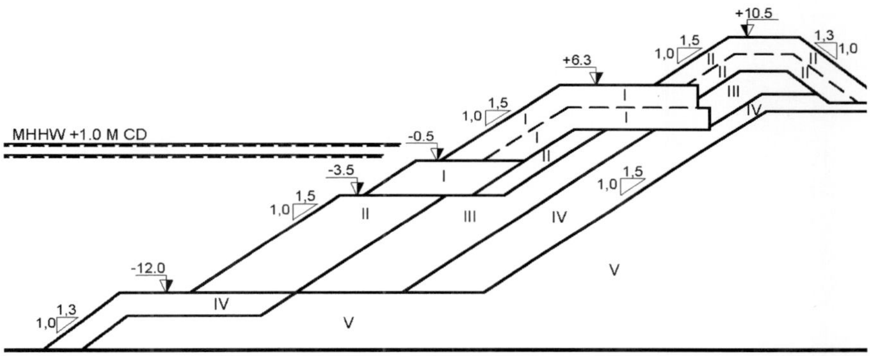

Figure 7.8. A sacrificial berm with Class I rock just below SWL gives support to the interlocked stones of Class I above it.

When placing armourstone above water level in two layers or in any other thickness it is essential that the operator of the excavator knows where the final surface should be. If the operator does not know this, it is likely that the final surface does not meet the required layer thickness or placement tolerances. The location of the final surface is often visualised for the operator by templates and sometimes there is an instructor assisting the operator.

More modern and now widely used by contractors is a GPS system coupled with a 3D model of all layers of the breakwater displayed in front of the operator. Using this technique together with the placement method described above, where the outermost stone at each level is placed first before filling stones back of it, has proven to result in good workmanship and aesthetically better placement. It also overcomes the problem of placing more rock than necessary with positive tolerances.

Two examples of such placement are given in Figure 7.9 for the extension of the Helguvik breakwater and in Figure 7.10 for the western breakwater head at Landeyjarhöfn.

Figure 7.9. Extension of Helguvik with special placing of Class I rock.

Figure 7.10. Western breakwater head at Landeyjahöfn with special placement of Class I rock.

7.4.2 *Tolerances*

As stated previously, berm breakwaters are mostly constructed of bulk-placed armourstone. This generally applies to all material below low water level, perhaps -0.5 m or -1.0 m, and all material inside the berm. Only the primary armour on the surface of the Icelandic-type breakwater is constructed as placed armour. Tolerances for bulk-placed armourstone, similar to those advocated in the Rock Manual [2007], may therefore apply to this material which are less strict than those for placed armour.

The proposed placing tolerances for bulk-placed armourstone and core are detailed in Table 7.2, where tolerances are measured perpendicular to the design lines.

Table 7.2. Placing tolerances for bulk-placed armourstone and core for berm breakwaters.

Level	*Design profile to constructed profile for Armourstone classes*	*Design profile to constructed profile for Core classes*
Above 0 m CD	+0.4 m to -0.2 m	+0.2 m to -0.2 m
0 m to -5 m CD	+0.8 m to -0.3 m	+0.5 m to -0.3 m
-5 m to -15 m CD	+1.2 m to -0.4 m	+1.2 m to -0.4 m
Below -15 m CD	The greater of: +1.5 m to -0.5 m or +1.5D_{n50} to -0.5D_{n50}	+1.5 m to -0.5 m

Placing tolerances for placed primary armour shall be in accordance with Table 7.3, where tolerances are measured perpendicular to the design lines. Notwithstanding the given tolerances, the following criteria shall apply to the armourstone layer:

- All tolerances refer to design and constructed profiles as appropriate, unless stated otherwise. Definition of constructed profile is given in Section 7.4.3;
- The tolerance on two consecutive constructed profiles shall not be negative;
- Notwithstanding any accumulation of positive tolerances on underlying layers, the thickness of the armour layer shall not be less than 90 percent of the designed thickness when calculated using constructed profiles.

Table 7.3. Placing tolerances for placed primary armour on berm breakwaters.

Level	*Individual measurements*	*Design profile to constructed profile*
Above -0.5 m CD	+/- 0.3 D_{n50}	+0.35 to -0.25 D_{n50}
Below -0.5 m CD	+/- 0.5 D_{n50}	+0.6 to -0.4 D_{n50}

7.4.3 *Definition of rock surface and survey method for constructed profile*

When contracts for breakwater construction are based on volume, the definition of the rock surface becomes an important issue. Figure 7.11 to Figure 7.13, from recently constructed berm breakwaters using extra-large armourstone with rough placement, clearly demonstrate the need to have a clear definition of the rock surface.

Figure 7.11. Narrowly graded stones on the Husavik berm breakwater, Iceland. Classes I and II, 16-30 t and 10-16 t.

In the North Atlantic the rock surface of rubble mound structures has for many years been defined as the plane through which armourstones protrude by one third of the surface area. This is an easy definition to place in the specifications, but more difficult to control.

Figure 7.12. The Sirevåg berm breakwater, Norway, final inspection of the 20–30 t Class I on the berm.

Figure 7.13. Class I stones, 20-35 t on top of the Icelandic-type berm of the breakwater protecting the Hammerfest LNG plant, Norway.

In some projects, the constructed surface was checked with a detailed survey but often this was done with the survey rod, not placed on the highest points of the stones, but reasonably low for a subjective evaluation of the rock surface definition. In other projects it was agreed-upon by the contractor and supervisor to base the definition of the constructed profile or surface on a highest point survey, lowered by either a fixed distance or a distance calculated as a factor times the nominal diameter of the armourstone class. Here it is recommended to develop the last mentioned method and to use the modern GPS staff survey equipment.

The Rock Manual [2007] advocates that rock surfaces should be surveyed with a spherical foot staff with a diameter of $0.5D_{n50}$, also called a survey ball. The idea behind this survey method is that the staff is neither placed on top of stones nor at low levels or between stones and it results in a surface that is not far from the definition used in the North Atlantic. But the operation, measuring with the spherical foot staff, is both expensive, time-consuming and takes up valuable space on the breakwater, usually requiring a crane and involving several persons. Therefore it is proposed to use the more modern GPS rod survey instead, where applicable.

It is recommended that when above water, the constructed profile shall be determined by measuring the highest point of armourstones with a GPS rod (see Figure 7.14 and Figure 7.15). The constructed profile shall be defined as a factor times the nominal stone diameter, D_{n50}, beneath the measurements. This factor will depend on the armourstone shape and placement, and may be determined from a test panel and in agreement with the client's representative.

Survey profiles shall be taken at 10 m intervals and set out so that they are perpendicular to the set out lines as defined on the construction drawings. Survey profiles may be based on highest point measurements within $1D_{n50}$ to each side of the measured profile. The interval between measurements shall be the greater of 1 m and $0.75D_{n50}$ for the armourstone class layer being surveyed.

Survey of the underwater profile shall be carried out using a multi-beam echo sounder. Determination of the constructed profile from underwater surveys may be based on highest point measurements and in agreement with the client's representative. Profiles from the underwater survey shall be connected to the above water survey on drawings. Where the tidal range is small, or wave agitation is substantial, there is often a gap between above and underwater surveys. One possibility to close this gap is to use the GPS 3D model machine control system of the excavators as a measuring device. The contractor shall be required to make every effort to close or minimise the gap between above water and underwater survey.

Figure 7.14. Surveying of Class I stones with a GPS staff, 16–30 t, on top of the berm before building of the crest section, Husavik berm breakwater, Iceland.

Figure 7.15. The Hammerfest breakwater, Norway, inspection survey with a GPS staff of the 20–35 t Class I stone of the unfinished berm.

7.5 Weight assessment of rock gradings *in-situ*

As a part of the breakwater construction, armourstone are sorted into classes according to weight. This is often done in the quarry and sometimes when extracting stones from a stockpile. Earlier, sorting by weight was mainly done by a visual evaluation of the operators of excavators or front loaders used for the sorting. Weighed and labelled stones of minimum, median and maximum weight of each class were placed in sight of the sorting operators. Nowadays it is more common that excavators and font loaders are equipped with load cells that weigh and register all larger armourstone.

Sometimes it is necessary to evaluate weight of armourstone *in-situ*. This could be a part of supervision, a monitoring programme or evaluations of damages. In this case, weighing of the placed armourstone is not possible. Two options will be described and compared. The first one is to measure "average" rock dimensions, the "three-side method" in Section 7.5.1.; and the other method is the assessment based on block shape and dimensions in Section 7.5.2.

7.5.1 *Weight assessment by estimating average rock dimensions*

If the three "average" measures of length l_{av}, thickness b_{av} and height h_{av} of an individual rock have been measured and the mass density of the rock has been established, then the estimation of the rock mass is as follows:

$$M = \rho_r.l_{av}.b_{av}.h_{av} \qquad 7.1$$

The main question is, however, how good the estimation of the "average" measures was, as a rock has quite an irregular shape. It may help if the method that is performed by a certain individual can be validated against a few examples with known weights. An example will be given here.

At a certain construction site the rock gradings from the quarry were established "by eye", using a few rock sizes with known weights close to the stock pile (see Figure 7.16). The requested grading was 1.5-3 t. First a number of stones were measured by the "three-side method" and calibrated against the known mass. This calibration is specific for the individual performing the measurements. It turned out that the average measures that were taken were too low and the calculated weights in average were a factor of 0.74 too low (see Figure 7.17). The measurement system, however, was not changed and also applied for the breakwater armour layer that had already been constructed (see Figure 7.18).

Figure 7.16. Calibration of "three-side method" on known rock masses from 1-6 t.

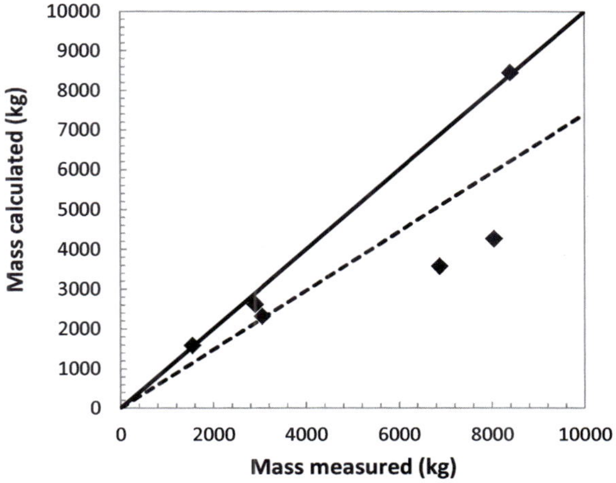

Figure 7.17. Calibration of the "three-side method", giving an underestimation of a factor 0.74.

Figure 7.18. Measuring rock sides on the constructed armour layer.

Rock weights measured at the armour layer were adjusted by the factor 0.74, established from the calibration. The final outcome is a grading curve given in Figure 7.19. The grading curve established by eye resulted in a much wider and larger grading than required. The grading should be 1.5–3 t, where the actual grading was closer to 1–5 t, which is a large deviation. The same could be applied to the underlayer rock that was required as 100–300 kg, but in reality was closer to 100–500 kg rock.

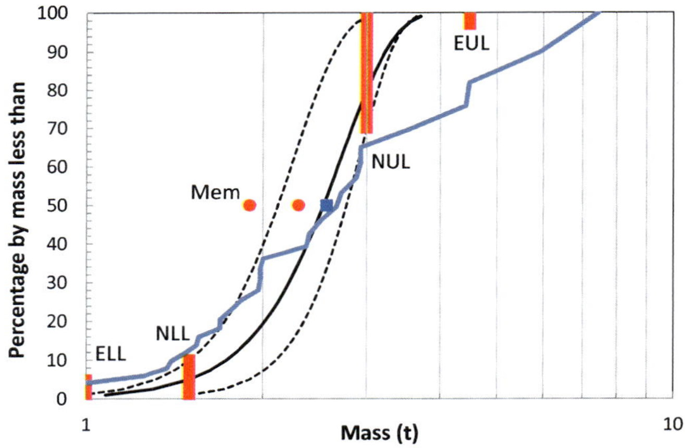

Figure 7.19. Grading curve established by the "three-side method", compared with the required grading of 1–3 t.

It may be concluded that producing grading curves "by eye" is not a very successful method.

7.5.2 *Weight assessment of rock in-situ from block shape*

The blockiness coefficient developed by Stewart *et al.* [2003] appeared to be a valuable method to describe the weight of a rock by its shape. The blockiness coefficient BL_c is defined as the ratio between the actual volume of a rock and the smallest rectangular box which fits the rock:

$$BL_c = V_{rock}/(l.b.h) \qquad 7.2$$

where:

BL_c	= blockiness coefficient (-)
V_{rock}	= volume of the rock (m^3)
l	= length, largest dimension of the "box" (m)
b	= width of the "box" (m), middle dimension
h	= height of the "box"(m), smallest dimension

The values of l, b and h can fairly easy be measured *in-situ*. Also the mass density of the rock ρ_r is easy to measure *in-situ*. If the blockiness coefficient, or shape factor, would be known for a certain rock, the mass M would be known without weighing the rock itself:

$$M = \rho_r.BL_c.l.b.h \qquad 7.3$$

This idea was developed further in Van der Meer *et al.* [2005] and Nurmohamed *et al.* [2006] and the final method can be used for rock structures or breakwaters to get a good measure of the actual grading of the rock on a built structure. First of all small rock samples with different shapes were gathered. All these small rocks were ranked according to their measured BL_c-value and photographs of this ranking were made. Figure 7.20 shows a picture of 25 rocks in a ranking, where the range found was BL_c = 0.33–0.68. Other armour units for coastal protection were taken and also the blockiness coefficients of these units were measured, just for comparison with rock. Figure 7.21 gives an overall view.

Figure 7.20. Different rock shapes ranked according to their blockiness coefficient BL_c.

The photograph in Figure 7.20 can be used to assess the actual blockiness coefficient in reality on a coastal protection or armour layer of a breakwater. It should be noted that $BL_c = 1$ is according to a cube and a value of 0.52 describes a sphere. It appears that nice blocky shapes have the highest coefficients around $BL_c = 0.6–0.7$. Irregular shaped rock with triangular shapes have lowest values with $BL_c = 0.3-0.4$. More rounded shapes come close to a ball or sphere with values around $BL_c = 0.5$.

Figure 7.21. Blockiness coefficients for armour units.

The method was validated with natural rock in a stockpile. This method can be applied to rock armour protections if the rock size and grading are not known. An example is given below.

Various locations along the Dutch coast with rock armour were analysed in detail. Each time the three dimensions of ten rocks were measured. The weight of each rock was predicted based on the guessed blockiness coefficient of each rock. The ten rocks together give a (rough) grading curve as shown in Figure 7.22. In this figure the grading curves of all measurement locations are given. Table 7.4 gives a summary of the findings, such as the average mass M_{50}, the nominal diameter D_{n50}, the grading D_{n85}/D_{n15} and the average blockiness coefficient. The average mass was not established by reading the value on the 50%-line, but by estimating a straight line through the grading curve, matching the trend of the grading between the 15%- and 85%-lines. Therefore, values of D_{n85}/D_{n15} in Table 7.4 differ from direct readings from the graph.

The average mass of the rock, established in this way, varies between 100-350 kg with corresponding nominal diameters of 0.35 to 0.50 m. The gradations of the rock are between D_{n85}/D_{n15} = 1.2 and 1.7, which is a normal gradation for quarried rock in coastal protections. The blockiness coefficients in average were about BL_c = 0.5.

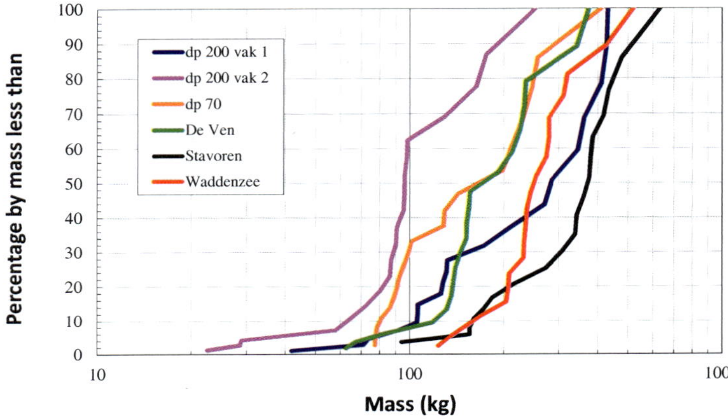

Figure 7.22. Grading curves for various rock armoured slopes, based on a non-intrusive method to establish the weight of a rock

Table 7.4. Measured rock grading at various locations along the Dutch coast.

Location	M_{50} [kg]	D_{n50} [m]	D_{n85}/D_{n10} [-]	Bl_c [-]
dp 200 vak 1	233	0,45	1,66	0,50
dp 200 vak 2	107	0,34	1,26	0,50
dp 70	157	0,39	1,52	0,49
De Ven	183	0,41	1,33	0,53
Stavoren	348	0,51	1,29	0.57
Waddensea dike	254	0,46	1,18	0,52

7.5.3 *Comparison of the two methods*

In order to compare the two methods described in the previous sections, a number of ten stones that were used for model testing were measured. The range in stone size was from 91 g to 413 g and the stones are shown in Figure 7.23. Two persons measured the rock according to average sizes as described in Section 7.5.1. One person measured according to maximum sizes and the blockiness coefficient BL_c, as described in Section 7.5.2.

Figure 7.23. Ten stones used to estimate the mass by measuring sides.

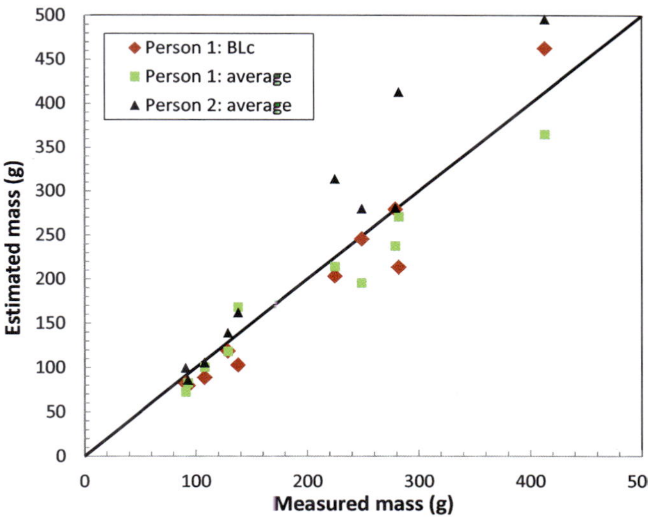

Figure 7.24. Ten stones measured by average size and by estimating the blockiness coefficient BL_c.

The results are shown in Figure 7.24 where the estimated mass is given versus the real mass (measured mass). Person 1 had a clear tendency to under estimate the mass, while person 2 did the opposite. The bias for person 1 by using BL_c was -10% and by using the average sizes it became -8%, which is consistent. The bias for person 2 by using the average sizes, however was +14%.

In fact, the comparison is not decisive in concluding which method is best. Both methods can be used, where probably the method of using average sizes is the easiest as one does not have to estimate the blockiness coefficient for each stone. The most important message is that measuring stones in the field should be calibrated by also measuring a number of stones with known weights. Any bias can then be taken into account.

Chapter 8

Geometrical Design into Practice, Examples

8.1 Design methods

8.1.1 *Geometrical design method for berm breakwaters*

The analysis and resulting design formulae in Chapters 3 and 4, together with the geometrical design rules developed in Chapter 5, and finally the practical experience described in Chapters 6 and 7, lead to a quite straight forward method to design berm breakwaters. The behaviour of the berm breakwater under design conditions depends very much on the type of berm breakwater that is considered: hardly, partly or fully reshaping.

There is a division in Icelandic-type berm breakwaters, with often 3–4 large rock classes in the berm, and the mass-armoured berm breakwaters. These have originally only one wide rock class for the berm, but as stated in Section 5.2.5 and Figure 5.11 it is strongly advised to divide the berm in two narrower rock classes. This means that such a mass-armoured berm breakwater comes close to an Icelandic-type berm breakwater with only two classes. Table 8.1 gives the classification of berm breakwaters with their expected behaviour and is equal to Table 2.2.

Figure 8.1 shows the principal cross-section of an Icelandic-type berm breakwater with the main geometrical design parameters as described in Chapter 5. With only two rock classes in the berm, it becomes the (newly proposed) mass-armoured breakwater.

Table 8.1. Classification of berm breakwaters based on the 100-years condition.

Type of breakwater	*Abbrevation*	$H_{sD}/\Delta D_{n50}$	S_d	Rec/D_{n50}
Hardly reshaping berm breakwater (Icelandic-type)	**HR-IC**	1.7-2.0	2-8	0.5-2
Partly reshaping Icelandic-type berm breakwater	**PR-IC**	2.0-2.5	10-20	1-5
Partly reshaping mass-armoured berm breakwater	**PR-MA**	2.0-2.5	10-20	1-5
Reshaping berm breakwater (mass-armoured)	**FR-MA**	2.5-3.0	--	3-10

The geometrical design means that the parameters in Figure 8.1 have to be established. As most of them have been given in formulae, it is quite easy to make a spreadsheet and calculate the parameters automatically. Design choices can be then made and a first cross-section can then be drawn, also in the same spreadsheet. The outcome may need further modification and will finally result in the design drawing of the breakwater cross-section.

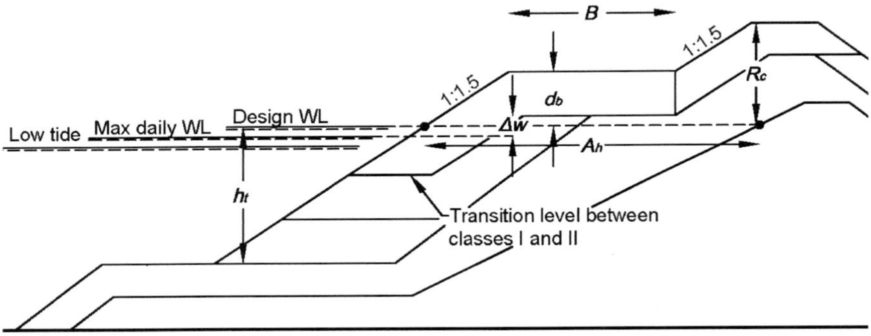

Figure 8.1. Principal cross-section of an Icelandic-type berm breakwater with the main geometrical design parameters.

The developed spreadsheet will be described here in depth and will then be used for all examples given in Sections 8.2 to 8.4, without further explanation of the spreadsheet. The first part gives the general design conditions as given in Table 8.2. A grey cell means that it is a requirement to give this item a (design) value. Design wave heights have to be given, the H_{sD} for the 100-years return period and the overload condition, which could be about 20% higher, or based on a much longer return period. In this way a performance-based design is achieved. Design water levels have to be given, as well as the wave height associated with a low water level. Finally, allowable overtopping has to be given for both design conditions, as well as mass densities of rock and water.

Table 8.2. General conditions in the design spread sheet. Grey cells are inputs required by the user.

General conditions			*Remarks*
Design wave height H_{sD}	5	m	100-years return period
Peak period T_p	10.3	s	
Overload H_s	6	m	About 1.2 H_{sD}
Design water level DWL	2	m CD	
Lowest water level with H_{sD}	1	m CD	
Lowest storm level	0	m CD	
H_s at lowest storm level	4.5	m	
Mean High Water Spring	1	m CD	
Bottom level of foreshore at toe of structure	-10	m CD	
Allowable overtopping q for H_{sD}	1	l/s per m	
Allowable overtopping q for overload	10	l/s per m	
Mass density water	1025	kg/m^3	
Mass density rock	2700	kg/m^3	

The next item is the specification of rock classes that are foreseen for the berm of the breakwater. After first calculations or a changed quarry yield prediction or output, it might be possible that this input will change and lead to a slightly adjusted cross-section. Two rock classes should be specified for a mass-armoured berm breakwater, leaving a blank in the two lowest cells, and three classes for an Icelandic-type berm breakwater. One may give the 0% and 100% class limits, as often done for the Icelandic-type designs, but also the limits as given for the standard heavy gradings in the Rock Manual [2007]. These are

NLL < 10% and NUL > 70%, where NLL and NUL describe the class boundaries.

Table 8.3. Rock classes to be specified in the design spreadsheet.

Choice of rock classes		
Rock Class I: minimum mass (0-10%)	5	t
Rock Class I: maximum mass (70-100%)	10	t
Rock Class II: minimum mass (0-10%)	1	t
Rock Class II: maximum mass (70-100%)	5	t
Rock Class III: M_{min} (leave open for MA)	0.3	t
Rock Class III: M_{max} (leave open for MA)	1	t

Table 8.4. Outcome of main parameters in the design spreadsheet.

Outcome of main parameters			*Remarks*
Wave steepness s_{op}	0.030	-	
Relative mass density Δ	1.63	-	
Median mass Class I M_{50}	7.5	t	Middle of the class limits
Nominal diameter Class I D_{n50}	1.41	m	
Stability number $H_{sD}/\Delta D_{n50}$	2.18	-	
Type of berm breakwater	Partly reshaping		Table 8.1
Number of rock classes for berm	3		
Basic recession for H_{sD} (no adaptation)	3.38	m	Equation 3.19
Recession for overload (no adaptation)	6.08	m	Equation 3.20
Nominal diameter Class II, D_{n50}	1.04	m	Middle of the class limits
Nominal diameter Class III, D_{n50}	0.62	m	Middle of the class limits

All data specified in Table 8.2 and Table 8.3 lead to the first outcome, mainly on related parameters and the type of berm breakwater. An overall view is given in Table 8.4. The wave steepness has been calculated, as well as M_{50}'s or D_{n50}'s, where these are based on the middle of the class limits. For example the class I 5–10 t armourstone has

M_{50} = 7.5 t, with an associated D_{n50} = 1.41 m. This leads to a stability number $H_s/\Delta D_{n50}$ for design conditions and consequently to the characteristics of the berm breakwater: hardly, partly or fully reshaping (based on Table 8.1).

The basic recession has been calculated, based on Equation 3.19 for the design condition and on Equation 3.20 for the overload condition. This is a basic recession as not all possible influences on recession, positive as well as negative, as described in Table 3.5 have yet been taken into account.

The berm width of a berm breakwater is closely related to the expected recession, but even more on the wanted resiliency, see Table 8.5. Small expected recession gives a large resiliency and vice versa. It is a designer's explicit choice how resilient he or she wants to design the berm breakwater. There is a relationship between the expected recession and resiliency, as in Equations 5.1 to 5.3, but the resiliency is given as a range. This is 10-20% for a hardly reshaping berm breakwater, 20-40% for a partly reshaping and less than 70% for a fully reshaping berm breakwater. The percentage is the part of the berm that might be eroded by the design wave height H_{sD}. A measure for the berm width is then given by Equation 5.4, where Equation 5.5 gives a minimum berm width based on a required minimum number of stones (geometry).

Table 8.5. Berm width and level, based on resiliency, as in the design spreadsheet.

Resiliency, berm width and level			*Remarks*
Wanted resiliency	30	%	Equations 5.1 to 5.3
Resulting Berm width B from resiliency	11.26	m	Equation 5.4
Minimum berm width B_{min} from geometry	4.78	m	Equation 5.5
Berm level 0.6 H_{sD}	5	m CD	Equation 5.10
Δw for waves during construction	1	m	Safety measure
MHWS plus Δw = working level	2	m CD	
Minimum berm level from construction	4.81	m CD	Above level + 2 D_{n50} Class I
Design choice of berm width	12.00	m	
Design choice of berm level	5.00	m CD	

The berm level is a free choice, but a berm level above a certain value ($\geq 0.6\ H_{sD}$, see Equation 5.10) increases stability and reduces recession, see also Table 3.5. Therefore this level is given to base a final design choice on. In Table 8.5 this comes to a level of +5 m CD. A minimum berm width follows from the construction procedure if one wants to construct Class I rock by working on the top level of the Class II rock and take into account some safety with regard to frequent waves. This safety is given by Δw and must be related to expected frequent wave conditions during construction. This safety is put on top of the level for MHWS. This all results in a minimum berm level, in the example in Table 8.5 this is +4.81 m CD.

Based on the outcome in Table 8.5, the designer has to make two design choices: the berm width and the berm level. The final berm width may also depend on the application of Table 3.5 (positive and negative influences) and is chosen here as 12.0 m. The final berm level can be based on the two calculated levels and here the highest level is chosen to have the positive influence of a high berm level on recession: 5.0 m CD.

The horizontal armour width, A_h, determines more or less the volume of large rock in the berm, see Figure 8.1. The minimum horizontal armour width is given by Equation 5.9 and has been calculated in Table 8.6. The designer's choice in the table is quite close to this value.

Table 8.6. The horizontal armour with A_h in the design spreadsheet.

Required horizontal armour width A_h	21.8	m	Equation 5.9
Design choice of A_h	22.0	m	

The transition from Class I to Class II rock at the seaward slope should not be too high, as the Class II rock will result in more recession to the structure if this rock is attacked by waves. Section 5.2.8 gives some guidance. One could consider the lowest water level possible with the design wave height H_{sD}, as well as a lower water level with a little smaller wave height. For an Icelandic-type berm breakwater with at least three rock classes in the berm, the highest level of transition is 0.4 H_s below the water level considered and for a mass-armoured berm breakwater with two classes this may be 0.6 H_s. Table 8.7 gives the calculation and the designer has to make a final choice on the level.

Table 8.7. Transition level from Class I to Class II as in the design spreadsheet.

Minimum transition level to Class II		*Remarks*
For H_{sD} at lowest level	-1.0 m CD	Section 5.2.8
For lowest level with according H_s	-1.8 m CD	Section 5.2.8
Design choice of transition for IC (3 rock classes)	-1.8 m CD	
Transition lower class for MA (2 rock classes)	-2.7 m CD	Section 5.2.8

The crest level design of a berm breakwater depends very much on what wave overtopping should be allowed. Based on the analysis of tested berm breakwaters (Table 3.2) most berm breakwaters had a crest level 1.2 to 1.4 H_{sD} above design water level (Equation 5.6). These limits have first been calculated in Table 8.8. But if an allowable overtopping discharge has been given, the required influence factor γ_{BB} can be calculated and subsequently the required crest level by the design equation on wave overtopping, Equation 4.13. Perpendicular wave attack has been assumed.

Table 8.8. Crest level calculations as in the design spreadsheet.

Crest level ($\gamma_\beta = 1$)			*Remarks*
If no overtopping criteria, $R_{c\ min}$	8.0	m CD	Equation 5.6
If no overtopping criteria, $R_{c\ max}$	9.0	m CD	Equation 5.6
For given allowable overtopping, q, γ_{BB}	0.42		Equations 4.8 and 4.10
Required crest level for design conditions	9.92	m CD	Equation 4.13
Required crest level for overload	9.64	m CD	Equation 4.13
Design choice of crest level	10.00	m CD	

Equation 4.13 gives a *design approach* with some safety on the outcome (about one standard deviation more). This safe design is proposed as prediction of wave overtopping is quite uncertain. The influence factor γ_{BB} is different for partly and hardly reshaping berm

breakwaters (Equation 4.10) and fully reshaping berm breakwaters (Equation 4.8). The official transition between the types is at $H_s/\Delta D_{n50} = 2.5$ and by using the two equations gives a discontinuity at this transition. In order to avoid a discontinuity, the crest levels in Table 8.8 have been calculated with Equation 4.10 if $H_s/\Delta D_{n50} < 2.3$ and with Equation 4.8 if $H_s/\Delta D_{n50} > 2.6$. For $2.3 \leq H_s/\Delta D_{n50} \leq 2.6$ both equations have been used and the crest levels have been interpolated. At the lowest row in Table 8.8, the designer has to make a choice on the crest level.

Figure 8.1 shows a foundation level for the large rock classes of an Icelandic-type berm breakwater, well above the level of the seabed. In the graph this is given as Class IV on a horizontal layer. For relatively deep water as well as for depth-limited conditions it is favourable for the stability of the berm if this foundation level is as high as possible, see Table 3.5. In principle one could consider this structure as a toe berm structure for a conventional breakwater and use toe stability formulae to assess the stability (Equation 5.11). A recently published alternative is the formula on toe rock stability of Van Gent and Van der Werf [2014]. In Table 8.9, Equation 5.11 has been used and the validity ranges are checked.

A check should be made whether the level of the designed toe can indeed be constructed (Figure 8.1). It is assumed that the core extends seaward with at least a thickness of 1.5 m. Then on top of this core the rock layer of the toe berm will be constructed, which has a thickness of at least 2 D_{n50}. The lowest level of the toe berm is then 1.5 m + 2 D_{n50} above the foreshore. These kinds of calculations have been performed in Table 8.9, for the design wave height as well as for the overload condition, where the designer has to give the allowable damage level for the toe for both conditions.

Finally, the designer has to make a choice on whether a toe is feasible and what the level should be. Another choice to be made is whether the slope angle of the core should be 1:1.5 or a little gentler in order to save some of the large material in the berm.

Table 8.9. Check on possibility of a toe berm as in the design spreadsheet.

Check possibility of toe berm at level h_t		*Remarks*
Lowest possible toe level (two layers)	-7.26	m CD
Design conditions		
Allowable damage level for H_{sD}, N_{od}	2	-
Highest level of toe for H_{sD} with chosen N_{od}	-6.78	m CD
Check validity range h_t/D_{n50}	12.5	ok
Check validity range h_t/h	0.71	ok
Overload conditions		
Allowable damage level for overload, N_{od}	4	-
Highest level of toe for overload with chosen N_{od}	-7.19	m CD
Check validity range h_t/D_{n50}	13.2	ok
Check validity range h_t/h	0.74	ok
Design choice of toe berm level (0 if no berm)	-7.2	m CD
Design choice cotα core below A_h	2	-

The final outcome of the design spreadsheet is a draft cross-section with a summary of the design choices. For the calculations and choices made above, this information is given in Figure 8.2. It shows the predicted recession as well as the horizontal armour width A_h and a division between the three classes.

The Class I rock in Figure 8.2, as given as output of the spreadsheet, has always a thickness of $2D_{n50}$. In the final design stage it may change, for example if the recession is quite large and could reach the underlying Class II rock. In that case it is possible to extend the seaward layer thickness to $3D_{n50}$ or more. This will be the case when $H_s/\Delta D_{n50}$ will be close to or larger than 2.2.

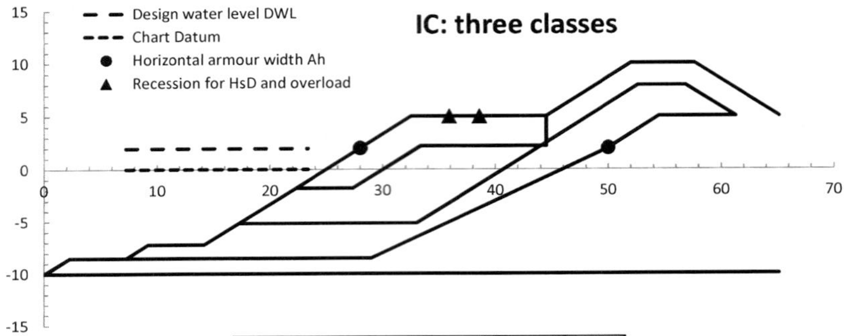

Summary of design choices	
Design of berm width	12.00 m
Design of berm level	5.00 m CD
Design of A_h	22.0 m
Design of transition class I to class	-1.8 m CD
Design of crest level	10.00 m CD
Design of toe berm level	-7.2 m CD
Design choice cotα core below A_h	2 -

Rock Class I	5 - 10 t
Rock Class II	1 - 5 t
Rock Class III	0.3 - 1 t

Figure 8.2. Draft cross-section and summary of design choices, based on the design spreadsheet.

8.1.2 *Design wave climate and other conditions for examples*

The geometrical design method described in Section 8.1.1 can be used to describe a number of examples. Examples of hardly, partly and fully reshaping will be given, as well as Icelandic-type berm breakwaters (at least three rock classes in the berm) and mass-armoured berm breakwaters (two classes of rock). As described in Chapter 6, there are differences between rock coming from a dedicated quarry and standard gradings as described in the Rock Manual [2007]. The maximum standard grading is 10–15 t, whereas it is often possible to get much larger rock, even classes like 20–35 t, from dedicated quarries.

The type of berm breakwater with its expected behaviour has been described in Table 2.2 and Table 8.1 and is mainly depending on the stability number. The kind of berm breakwater that can be designed depends further on the available (maximum) rock class and of course on the design wave height, H_{sD}. In order to come to useful examples some elementary data have been gathered in Table 8.10.

First possible rock classes have been given for a dedicated quarry, based on earlier experience (see also Chapter (6)) and then the heavy gradings as given in the Rock Manual [2007]. Three wave climates have been considered, a very moderate wave climate with H_{sD} = 3 m, a quite normal wave climate with H_{sD} = 5 m and an extreme wave climate with H_{sD} = 7 m. Table 8.10 gives the calculated stability numbers for each chosen rock class (except for underlayers) and for each wave climate, considering a mass density of the rock of 2700 kg/m^3 and of sea water 1025 kg/m^3. Actual stability numbers may change with other mass densities.

Table 8.10. Stability numbers for chosen rock classes and design waves. Grey cells give the examples described in this chapter.

		Stability number $H_{sD}/\Delta D_{n50}$		
Dedicated quarry	**M_{50} (t)**	**H_{sD} = 3 m**	**H_{sD} = 5 m**	**H_{sD} = 7 m**
Class 20-35 t	25.0	0.87	1.46	**2.04**
Class 10-20 t	15.0	1.04	**1.73**	**2.42**
Class 4-10 t	7.0	1.34	**2.23**	3.12
Class 1-4 t	2.5	**1.88**	3.14	4.39
Class 0.2-1 t	underlayer			
Class 2-6 t	4.0	1.61	**2.68**	3.76
Class 0.5-2 t	1.2	**2.41**	4.01	5.61
Standard gradings				
Class 10-15 t	12.5	1.10	**1.84**	**2.57**
Class 6-10 t	8.0	1.28	**2.13**	**2.98**
Class 3-6 t	4.5	1.55	**2.58**	3.61
Class 1-3 t	2.0	**2.03**	3.38	4.73
Class 0.3-1 t	underlayer			

Stability numbers smaller than $H_{sD}/\Delta D_{n50}$ < 1.7 mean that a hardly reshaping berm breakwater can (easily) be made. That is the case for most of the heavy rock classes and a design wave height of 3 m. For a wave height of 5 m, such small stability numbers can only be reached with a Class I rock of 20–35 t (giving $H_{sD}/\Delta D_{n50}$ = 1.46). This low stability number cannot be reached for a design wave height of 7 m. Stability numbers $H_{sD}/\Delta D_{n50} > 3.0$ mean that the structure would become dynamically stable and this is not acceptable for a berm breakwater. Mainly the smaller gradings and for design wave heights of 5 m and more show these large stability numbers.

The most interesting cases in Table 8.10 are stability numbers between $H_{sD}/\Delta D_{n50}$ 1.7 – 3.0. They give the area of the design of berm breakwaters. These numbers are bold figures in the table. But some numbers for the same design wave height are quite similar, mainly because some of the gradings for a dedicated quarry are quite similar to the heavy standard gradings in the Rock Manual [2007]. For this reason examples have been chosen from Table 8.10 in such a way that, if possible, all three types (hardly, partly and fully reshaping) are present for each design wave height and they are distributed over the rock classes for dedicated quarries as well as the heavy standard gradings. The cells with a gray colour are the examples that will be described in the next sections.

For a design wave height of only 3 m, it is not necessary to design a fully reshaping mass-armoured berm breakwater. Already with a Class I of 0.5–2 t it is possible to design a partly reshaping berm breakwater. In total eight examples have been chosen from Table 8.10.

8.2 Examples for a design wave height of 5 m

8.2.1 *HR IC dedicated quarry, Class I 10-20 t*

The design condition is a design wave height of H_{sD} = 5.0 m with a wave steepness of s_{op} = 0.03, giving T_p = 10.3 s. The overload condition is considered to be 20% higher than the design condition, giving H_s = 6.0 m at the design water level DWL. Tides range between 0 m CD to + 1 m CD (1 m tidal range). The design water level (100-years condition) = surge + maximum tide = +2 m CD. Daily waves are quite moderate and a safety margin of Δw = 1 m above MHWS will be enough for construction. A more or less flat foreshore is present and at the toe of the structure the bottom is present at -10 m CD. All input and calculated values, including design choices, are given in the spreadsheet in Appendix D.

The stability number is calculated as $H_s/\Delta D_{n50}$ = 1.73 for a Class I of 10–20 t, which indeed gives a hardly reshaping berm breakwater. The rock grading comes from a dedicated quarry. For this Class I, one can choose connecting classes as Class II = 4–10 t, Class III = 1–4 t and if necessary a Class IV = 0.2–1 t, all given in Table 8.10. Here three classes are chosen (see Figure 8.3).

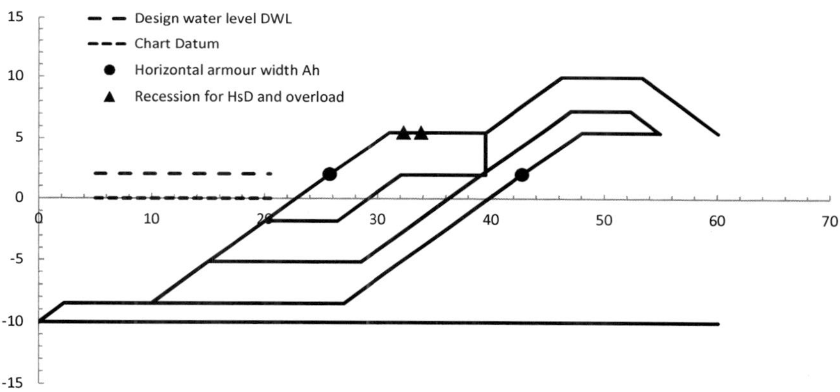

Summary of design choices	
Design of berm width	8.50 m
Design of berm level	5.50 m CD
Design of A_h	17.0 m
Design of transition class I to class I	-1.8 m CD
Design of crest level	10.00 m CD
Design of toe berm level	0 m CD
Design choice cotα core below A_h	1.5 -

Rock Class	10 - 20	t
Rock Class	4 - 10	t
Rock Class	1 - 4	t

Figure 8.3. Calculated cross-section for H_{sD} = 5 m and Class I = 10-20 t.

The wanted resiliency is taken at 15% reshaping. The allowable overtopping q = 1 l/s per m for the 100-years event and q = 10 l/s per m for the overload. This is quite strict for the 100-years condition and if the rock on the crest would be large enough; i.e. close to Class I, the allowable overtopping could be raised a little. Finally, the mass density seawater is $\rho_w = 1025$ kg/m^3 and of rock $\rho_r = 2700$ kg/m^3.

Figure 8.3 should be reviewed to come to a final design of the cross-section. In this chapter the experience with design and construction of berm breakwaters has been used to make this review. Figure 8.4 shows the final outcome and is of course quite close to the draft cross-section. The geometry of the Class I rock remains the same: a 3.5 m thick layer on the berm and down to -1.8 m CD at the seaward slope. In order to save a little Class II rock below the large Class I rock, a two-diameter thick layer was taken (2.7 m thick). Another measure was to change the slope of the core to 1:2, which uses less Class III rock.

The water depth is too small to design a higher toe berm and therefore only a layer of 1.5 m core extends as an apron.

In Figure 8.4, the breakwater protects a quay area. Therefore the core of the crest terminates at the quay level.

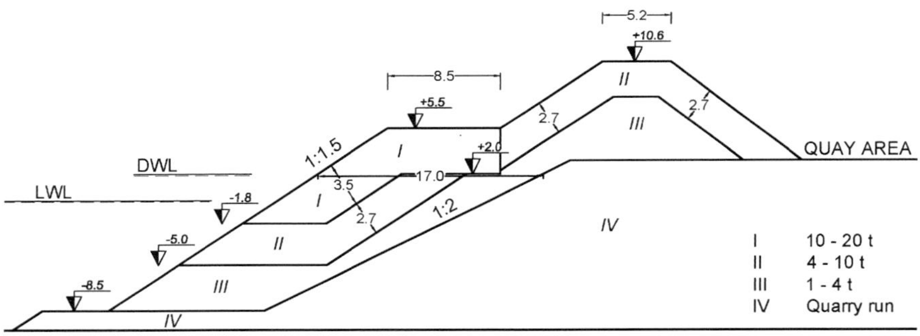

Figure 8.4. Hardly Reshaping Icelandic-type berm breakwater cross-section designed for $H_{sD} = 5.0$ m, Class I 10-20 t, $q_{100y} = 1$ l/s per m.

An allowable overtopping discharge of only 1 l/s per m in the 100-years design condition is quite strict. Figure 8.5 shows the lowering of the crest by about 2 m if 10 l/s per m could be tolerated. One should however realise that for the overload condition, the overtopping would increase to about 100 l/s per m, certainly significant overtopping.

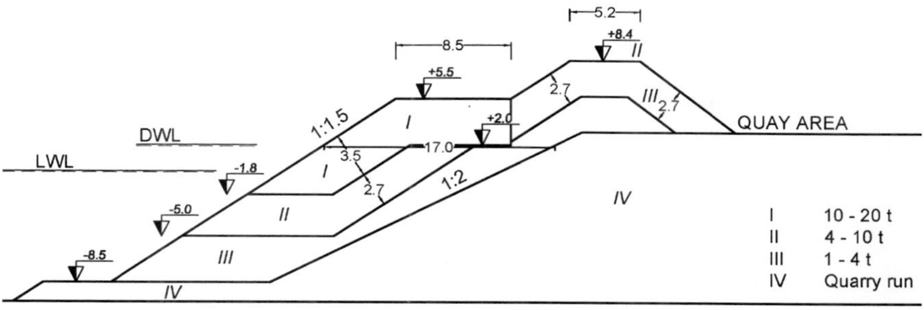

Figure 8.5. Hardly Reshaping Icelandic-type berm breakwater cross-section designed for $H_{sD} = 5.0$ m, Class I 10–20 t, $q_{100y} = 10$ l/s per m.

8.2.2 *PR IC standard gradings, Class I 6-10 t*

The hydraulic design conditions are similar to the ones in Section 8.2.1, and only the rock classes have been changed. A short summary of these conditions is given here. All input and calculated values, including design choices, are given in the spreadsheet in Appendix D.

$H_{sD} = 5.0$ m $\quad s_{op} = 0.03 \quad T_p = 10.3$ s $\quad$ Overload $H_s = 6.0$ m
Tidal range 0 m CD to + 1 m CD $\quad$ DWL = +2 m CD
Safety $\Delta w = 1$ m above + 1 m CD $\quad$ Seabed at -10 m CD

The stability number is calculated as $H_s/\Delta D_{n50} = 2.13$, for a Class I of 6–10 t, which indeed gives a partly reshaping berm breakwater. For this Class I from standard gradings, one can choose connecting classes also from standard gradings, as Class II = 3–6 t and Class III = 1–3 t, all given in Table 8.10. The three classes are given in Figure 8.6.

The wanted resiliency is taken at 30% reshaping, which is the middle of the proposed range. The allowable overtopping q = 1 l/s per m for the 100-years event and q = 10 l/s per m for the overload. Finally, the mass density of seawater is $\rho_w = 1025$ kg/m^3 and of rock $\rho_r = 2700$ kg/m^3, similar to the previous example. Figure 8.6 shows the outcome of the calculations.

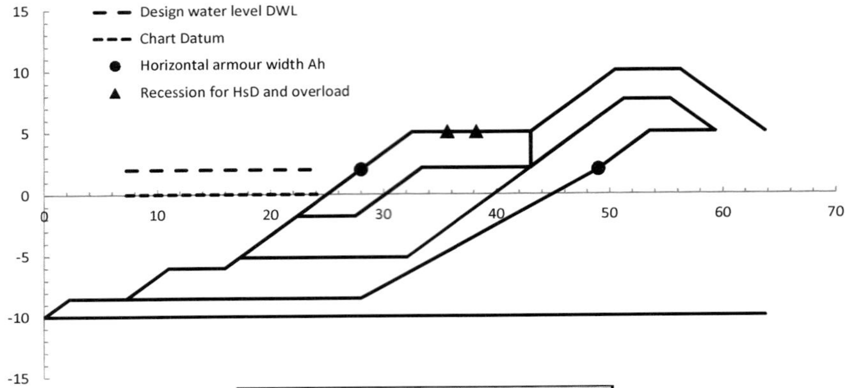

Summary of design choices	
Design of berm width	10.50 m
Design of berm level	5.00 m CD
Design of A_h	21.0 m
Design of transition class I to class I	-1.8 m CD
Design of crest level	10.00 m CD
Design of toe berm level	-6 m CD
Design choice cotα core below A_h	2 -

Rock Class	6 - 10	t
Rock Class	3 - 6	t
Rock Class	1 - 3	t

Figure 8.6. Calculated cross-section for H_{sD} = 5 m and Class I = 6–10 t.

The Class III rock of 1–3 t is still quite large compared to the wave conditions and this rock can be used to make a higher toe berm. The calculations (Appendix D) give a highest toe level of -5.16 m CD and a level of -6 m CD has been chosen. The transition of Class I to Class II rock on the seaward side was chosen at -1.8 m CD.

The expected reshaping in Figure 8.6 is given by the triangles, the left one by the design conditions, the right one by overload conditions. The erosion profile should start in these points and will show a S-shaped curve, which crosses the original lower slope just below the water level, see for example Figure 5.1 or Figure 5.2. The triangles in Figure 8.6 suggest that the erosion profile might also take a part of the Class II rock below, certainly for the overload conditions. For this reason the final design, as given in Figure 8.7, shows a seaward layer thickness of three stones, i.e. 4.3 m. Also the Class II layer underneath this layer has a thickness of three stones: 3.6 m.

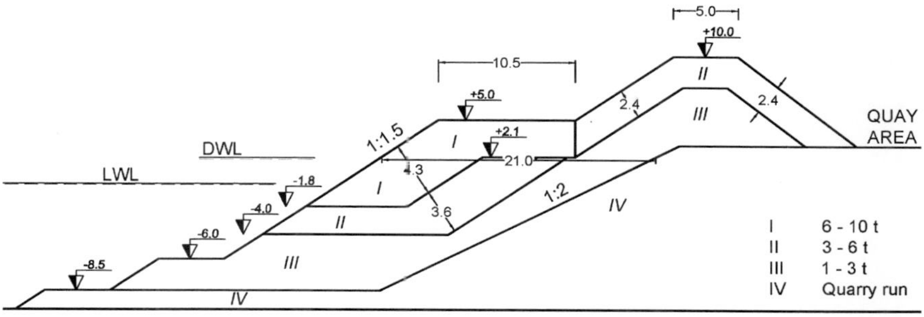

Figure 8.7. Partly Reshaping Icelandic-type berm breakwater cross-section designed for $H_{sD} = 5.0$ m, Class I 4-10 t, $q_{100y} = 1$ l/s per m.

8.2.3 *FR MA standard gradings, Class I 3-6 t*

The hydraulic boundary conditions in this example are similar to the ones in the two previous examples. A short summary of these conditions is repeated here. All input and calculated values, including design choices, are given in the spreadsheet in Appendix D.

$H_{sD} = 5.0$ m $\quad s_{op} = 0.03 \quad T_p = 10.3$ s $\quad$ Overload $H_s = 6.0$ m
Tidal range 0 m CD to + 1 m CD $\quad$ DWL = +2 m CD
Safety $\Delta w = 1$ m above + 1 m CD $\quad$ Seabed at -10 m CD

The stability number is calculated as $H_s/\Delta D_{n50} = 2.58$, for a Class I of 3–6 t, which indeed gives a fully reshaping berm breakwater. Originally a mass-armoured berm breakwater would have one large Class I rock with a wide gradation, for example 1–6 t. The stability number would then become $H_s/\Delta D_{n50} = 2.81$, still in the range of a fully reshaping berm breakwater, but with more recession of the berm. As proposed in Section 5.2.5 it is always an advantage to split the wide gradation into two gradings, in this case standard gradings of 3–6 t (Class I) and 1–3 t (Class II). The classes and cross-section are given in Figure 8.8.

The wanted resiliency is taken at 50% reshaping, which is 20% less than the maximum of 70%. The allowable overtopping q = 1 l/s per m for the 100-years event and q = 10 l/s per m for the overload. Finally, the mass density seawater is ρ_w = 1025 kg/m^3 and of rock ρ_r = 2700 kg/m^3, similar to the previous examples. Figure 8.8 shows the outcome of the calculations.

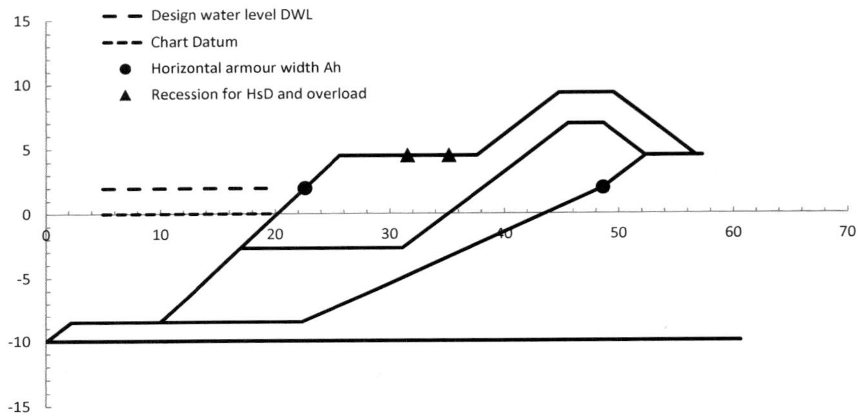

Summary of design choices	
Design of berm width	12.00 m
Design of berm level	4.50 m CD
Design of A_h	26.0 m
Design of transition class I to class I	-1.8 m CD
Design of crest level	9.30 m CD
Design of toe berm level	0.00 m CD
Design choice cotα core below A_h	2.50 -

Rock Class	3 - 6	t
Rock Class	1 - 3	t

Figure 8.8. Calculated cross-section for H_{sD} = 5 m and Class I = 3–6 t.

Figure 8.8 shows two details that need a better look. First of all, despite a resiliency of 50%, the recession of the overload condition (the righthand triangle in the graph) is quite close to the edge with the upper slope. This means that under overload conditions almost the complete berm would reshape. Secondly, the horizontal armour width of 26 m with a berm width of 12 m, gives a situation where the large rock extends far under the crest of the structure. Both observations lead to the conclusion that enlarging the berm width (or similarly: shifting the crest landward) would be a good option. Figure 8.9 shows the final design with a berm width of 14 m.

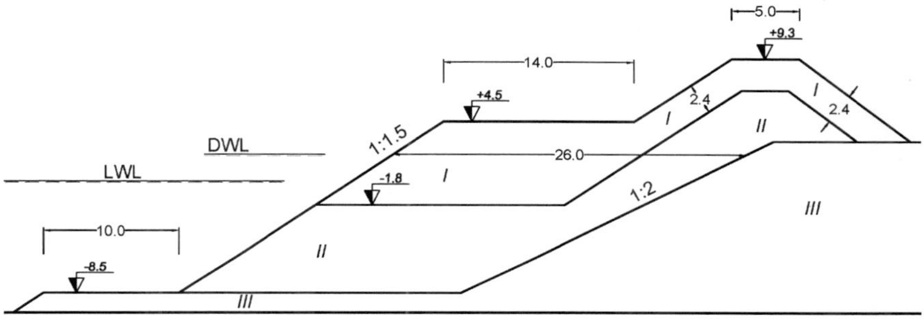

Figure 8.9. Fully Reshaping Mass-armoured berm breakwater cross-section designed for $H_{sD} = 5.0$ m, Class I 1–10 t, $q_{100y} = 1$ l/s per m.

8.2.4 *Conventional rock armour design*

With a design wave height of 5 m, as in the previous sections, and a fairly large rock grading like 6–10 t, it is also possible to design a conventional rock armour protection. Such a design can then be compared with, for instance, the partly reshaping Icelandic-type berm breakwater that is described in Section 8.2.2. That example has a Class I rock of 6–10 t.

The design conditions are given with a peak period of $T_p = 10.3$ s. Assuming a relationship of $T_p = 1.2T_m$, gives a mean period of $T_m = 8.6$ s. The peak of the storm is assumed to give 3000 waves, which is a little over 7 hours. Equations 3.1 to 3.3 can be used to calculate the damage level, S_d, for several wave conditions. Figure 8.10 gives the damage curves for three mean wave periods, as calculated by Breakwat.

The rock slope that was chosen was 1:2.5. A steeper slope would probably give too much damage. A gentler slope, like 1:3, would give less damage. Figure 8.10 also shows the design condition (100-years) and the overload condition. The design condition gives $S_d = 3.4$ and the overload condition gives $S_d = 6.8$. For a slope of 1:2.5 an allowable damage for a 100-years condition would be between $S_d = 2–4$. For an overload condition the damage should not exceed S_d = 10. Both conditions are met, which means that a slope 1:2.5 with a 6–10 t armour layer would be able to withstand the given wave conditions.

Figure 8.10 also shows the influence of the wave period. This influence is insignificant for berm recession of a berm breakwater, but is significant for a conventional armour layer of rock. A smaller wave period than the design period will clearly give less damage. But the damage increases quite rapidly if the mean period increases from T_m = 8.6 s to 10 s. The damage increases to S_d = 5.0 for the design condition and S_d =9.9 for the overload condition. For a good design of this conventional structure, one should look at the range of wave periods that will be possible. If T_m = 10 s would be possible for design conditions, one should modify the slope angle to 1:3 in order to make the structure more stable. In this example a slope of 1:2.5 is taken for design.

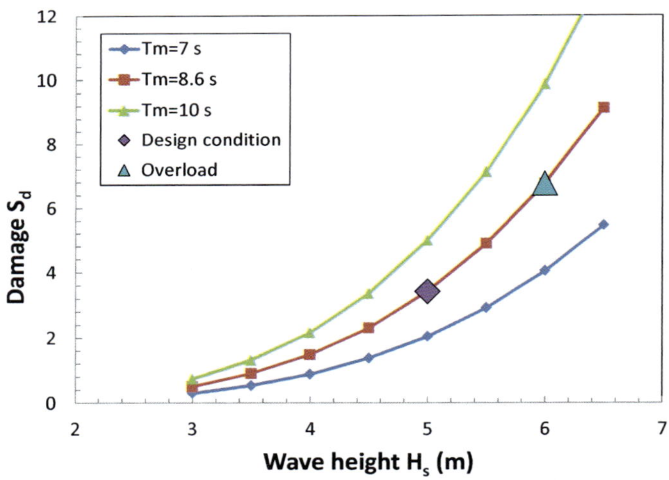

Figure 8.10. Damage curves for a conventional rock slope with an armour layer of 6–10 t; cotα = 2.5; P = 0.4; N = 3000.

The underlayer is normally 1/10th to 1/15th of the armour mass. This leads to a standard grading of 300–1000 kg. The rock armour layer has a nominal diameter of D_{n50} = 1.44 m, which gives a layer thickness of 2.9 m. The underlayer has a nominal diameter of D_{n50} = 0.62 m, which gives consequently a layer thickness of 1.25 m.

The required crest height can be calculated with the formulae in EurOtop [2016]. A conventional rock armour on an underlayer has a roughness factor of $\gamma_f = 0.40$. For the design conditions with $H_s = 5.0$ m and an allowable overtopping discharge of 1 l/s per m, a crest freeboard is required of 7.45 m. With a design water level of +2 m CD, the required crest height for this condition becomes 9.45 m CD. With the overload condition with $H_s = 6.0$ m and an allowable overtopping discharge of 10 l/s per m, the required crest freeboard becomes $R_c = 7.2$ m. This is lower than the 7.45 m for the design condition, which means that the crest height can be determined at 9.5 m CD.

The cross-section is given in Figure 8.11 and can be compared with the partly reshaping berm breakwater in Figure 8.7. Both structures are fit for purpose. A difference might be that the resiliency of the berm breakwater is larger than for the conventional structure, as after the overload condition still half of the berm is left, where the armour layer of the conventional structure will be close to "underlayer visible". Both structures can cope with such an overload condition.

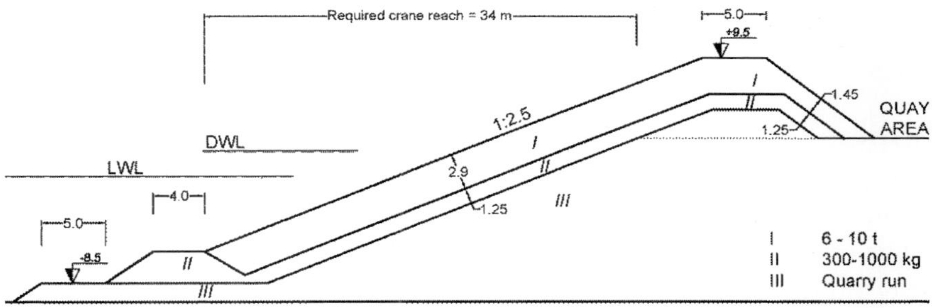

Figure 8.11. Conventional rock armoured structure with 6–10 t on a slope of 1:2.5. Design for $H_{sD} = 5$ m and $q_{100y} = 1$ l/s per m.

Figure 8.12 shows direct comparisons of the two cross-sections, where the conventional structure has been given in blue and the berm breakwater in red. The volume of large rock 6–10 t in the conventional design is more than twice of that in the berm breakwater (Class I). Of course the total volume of rock larger than 1 t is larger for the berm breakwater, but the use of the largest rock class is much smaller. In case the largest rock class is not easy to produce, the berm breakwater design gives an advantage. The total volumes of the cross-sections of both designs are quite similar, both around 460–480 m^3 per m length.

All rock on the berm breakwater of Figure 8.7 can be placed by excavator. This will be much more difficult for the conventional design, as the largest rock near the toe need a crane reach of about 34 m or placement from marine plant.

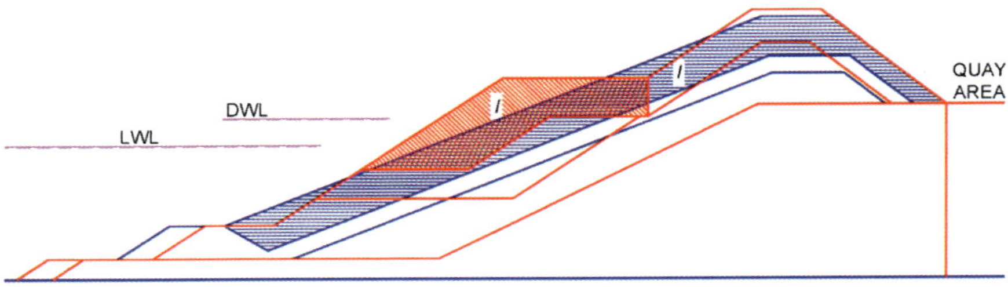

Figure 8.12. Direct comparison of cross-sections of the conventional breakwater and the berm breakwater.

8.2.5 *Overall conclusions and comparison*

Sections 8.2.1 to 8.2.4 give four designs of rock structures that can cope with a 100-years design wave height of 5 m. They are all fit for purpose. In all cases the crest level is around 10 m CD, allowing about 1 l/s per m wave overtopping. The main difference is the rock size of Class I on the seaward side and the volume of this largest rock class. A Class I of 10–20 t is of course a very large rock class, but the volume needed with respect to the total volume is quite limited, see Figure 8.4. The largest volume of rock for the breakwater is found for the fully reshaping berm breakwater, see Figure 8.9. However, the largest rock class is only 3–6 t. The difference between the berm breakwater designs is the resiliency. The smallest stability number, or largest rock size for Class I, gives the best resiliency. Even after overload conditions there is a large remaining capacity for severe wave action if the structure is only hardly reshaping. From that point of view one should always try to design for the largest rock class that can be made available.

The conventional structure in Figure 8.11 has a long 1:2.5 slope with 6–10 t rock. The armour layer cannot be constructed by excavator, but needs a crane with a long reach. Total rock usage is comparable with the berm breakwater with Class I rock of 6–10 t (Figure 8.7), but the need for the large 6–10 t rock is more than twice for the conventional design. This conventional design is more vulnerable for (longer) wave periods, which is not the case for a berm breakwater.

8.3 Examples for a design wave height of 3 m

8.3.1 *HR IC dedicated quarry, Class I 1–4 t*

The design condition is a design wave height of $H_{sD} = 3.0$ m with a fairly low wave steepness of $s_{op} = 0.02$, giving $T_p = 9.8$ s. The overload condition is 20% higher than the design condition, giving $H_s = 3.5$ m at the design water level DWL. Tides range between 0 m CD to + 1 m CD (1 m tidal range). The design water level (100-years condition) is at the same level as the maximum tide = +1 m CD. Daily waves are quite moderate and a safety margin of $\Delta w = 1$ m above MHWS will be enough for construction. A more or less flat foreshore is present and at the toe of the structure the seabed is present at -9 m CD. All input and calculated values, including design choices, are given in the spreadsheet in Appendix D.

The allowable overtopping q = 5 l/s per m for the 100-years event and q = 20 l/s per m for the overload. Finally, the mass density seawater is $\rho_w = 1025$ kg/m^3 and of rock $\rho_r = 2600$ kg/m^3.

Note that design conditions described above, the allowable overtopping conditions, as well as the mass densities of sea water and rock may differ from the conditions assumed in Section 8.2. On one hand this makes direct comparison between the solutions in Sections 8.2 to 8.4 not straight forward (within the section they are comparable), but on the other hand it shows reality, where these conditions vary.

The wanted resiliency is taken at 20% reshaping, which is the upper value of the proposed range. The stability number is calculated as $H_s/\Delta D_{n50} = 1.98$, for a class I of 1–4 t, which indeed gives a hardly reshaping berm breakwater. For this class I from a dedicated quarry, one can choose Class II = 0.2–1 t, both classes are given in Table 8.10. Only two rock classes are needed for this mild wave climate and the calculated cross-section is given in Figure 8.13.

The "standard" cross-section from the spreadsheet gives a berm breakwater with three rock classes, as in Figure 8.13. Actually, one should only look at the Class I layer, as the underlying material in the berm will all be Class II 0.2–1 t rock. A high toe berm has no function for a hardly reshaping berm breakwater and therefore a high toe berm has not been designed.

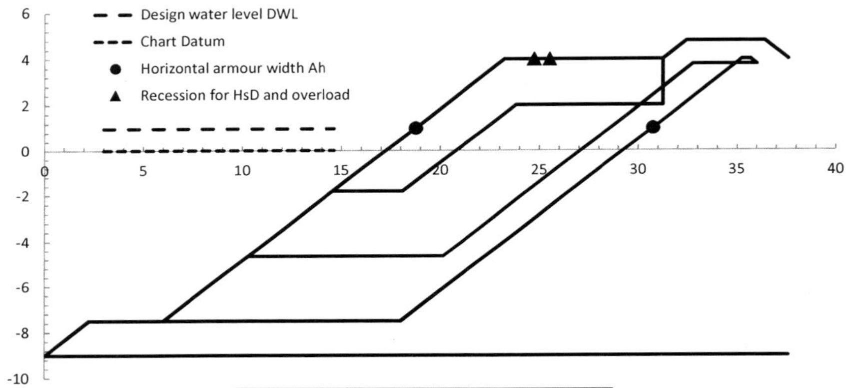

Summary of design choices	
Design of berm width	8.00 m
Design of berm level	4.00 m CD
Design of A_h	12.0 m
Design of transition class I to class I	-1.8 m CD
Design of crest level	4.80 m CD
Design of toe berm level	0.00 m CD
Design choice cotα core below A_h	1.50 -

Rock Class	1 - 4	t
Rock Class	0.2 - 1	t

Figure 8.13. Calculated cross-section for H_{sD} = 3 m and Class I = 1–4 t.

The berm level is higher than $0.6H_{sD}$, as the berm level follows from the construction issue that it should be constructed from the underlying Class II rock on a safe working level. But as the Class I rock is not heavy and distances to reach not very far, it may also be possible to construct the Class I rock with an excavator from the core. In that case the berm level can be lowered to roughly +2.8 m CD, instead of +4.0 m CD as in the final design given in Figure 8.14. Another option is to raise the crest level a little and have less wave overtopping at marginal costs.

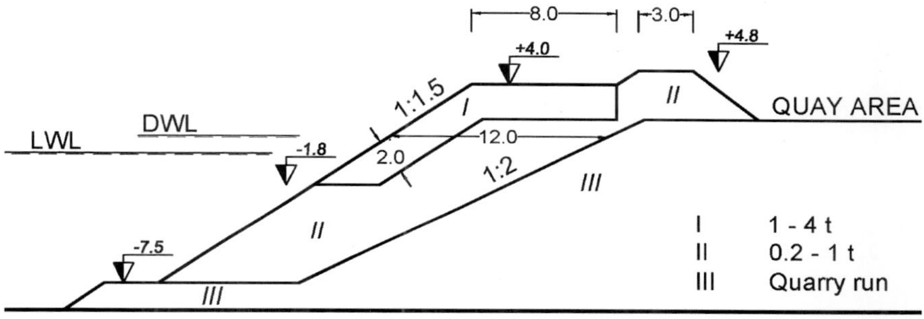

Figure 8.14. Hardly Reshaping Icelandic-type berm breakwater cross-section designed for H_{sD} = 3.0 m, Class I 1-4 t, q_{100y} = 5 l/s per m.

8.3.2 *PR MA dedicated quarry, Class I 0.5–2 t*

The wave conditions are similar to the example in the previous section. A short summary of these conditions is repeated here. All input and calculated values, including design choices, are given in the spreadsheet in Appendix D.

H_{sD} = 3.0 m s_{op} = 0.02 T_p = 9.8 s   Overload H_s = 3.5 m
Tidal range 0 m CD to + 1 m CD   DWL = +2 m CD
Safety Δw = 1 m above + 1 m CD   Seabed at -9 m CD

The design is a partly reshaping mass-armoured berm breakwater. The wanted resiliency is taken at 30% reshaping, which is the middle value of the proposed range. The stability number is calculated as $H_s/\Delta D_{n50}$ = 2.49, for a Class I of 0.5–2 t, which indeed gives a partly reshaping berm breakwater, but it is very close to fully reshaping. For Class I from a dedicated quarry, one can choose Class II = 100–500 kg,

both classes are given in Table 8.10. The mass-armoured breakwater has two rock classes and the calculated cross-section is given in Figure 8.15.

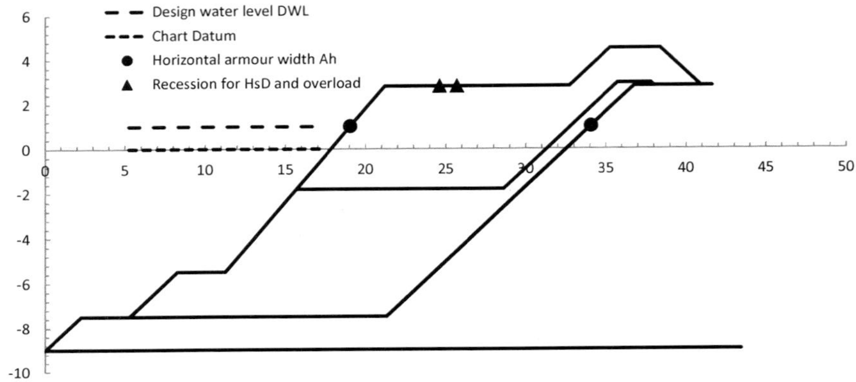

Summary of design choices	
Design of berm width	11.50 m
Design of berm level	2.80 m CD
Design of A_h	15.0 m
Design of transition class I to class I	-1.8 m CD
Design of crest level	4.50 m CD
Design of toe berm level	-5.50 m CD
Design choice cotα core below A_h	1.50 -

Rock Class	0.5 - 2	t
Rock Class	0.1 - 0.5	t

Figure 8.15. Calculated cross-section for H_{sD} = 3 m and Class I = 0.5–2 t.

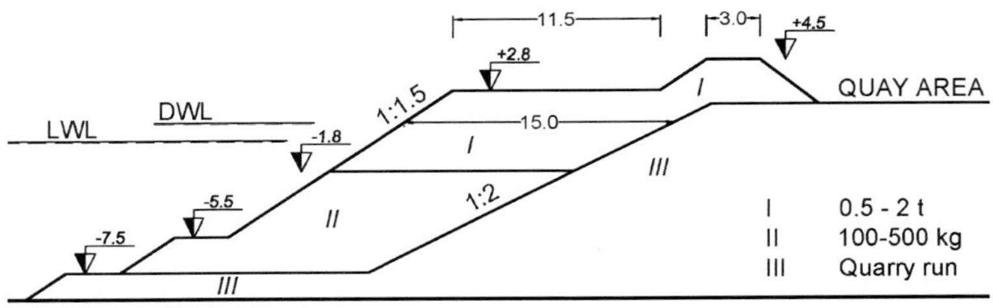

Figure 8.16. Partly Reshaping Mass-armoured berm breakwater cross-section designed for H_{sD} = 3.0 m, Class I 0.5–2 t, q_{100y} = 5 l/s per m.

A high toe berm has a positive function for a partly reshaping berm breakwater and therefore a high toe berm at -5.5 m CD has been designed. As for the design in the previous section, the berm level is higher than at $0.6H_{sD}$. The final design is given in Figure 8.16.

8.3.3 *Conventional rock armour design*

A design wave height of only 3 m is quite mild and a conventional rock armoured structure can certainly be designed without needing too large rock. Damage curves have been calculated in a similar way as in Section 8.2.4. First a rock armour of 1–4 t has been chosen, similar to the berm breakwater design in Section 8.3.1 (see Figure 8.17). A slope angle of 1:2.5 is needed for a stable design. This would result in a similar cross-section as in Figure 8.11, although with smaller rock.

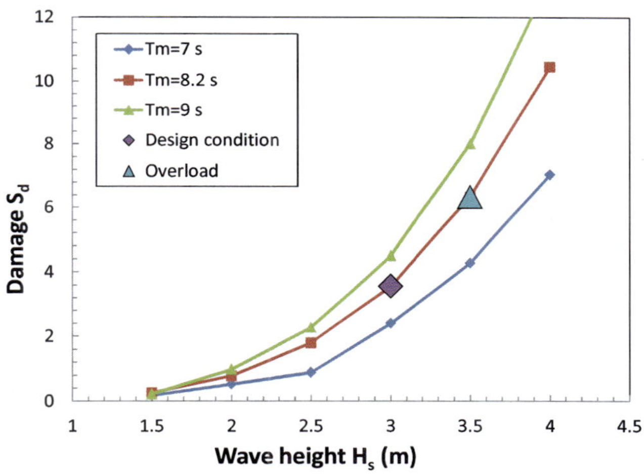

Figure 8.17. Damage curves for a conventional rock slope with an armour layer of 1–4 t; $\cot\alpha = 2.5$; $P = 0.4$; $N = 3000$.

A 1–4 t rock grading is not a very large grading. In order to overcome the problem with construction of a gentle slope (long crane or excavator reach) it is also possible to consider a slightly larger rock class, for example 3–6 t, which is also a standard grading.

Figure 8.18 gives similar damage curves as specified in Figure 8.17, but the rock grading is now 3–6 t and the slope angle required now becomes 1:1.75. This is almost as steep as the seaward slope of the berm breakwaters. An underlayer that will be acceptable, although a little on the small side with respect to the general rule of $1/10^{th}$ to $1/15^{th}$ of the armour layer mass, is 100–300 kg. A standard grading of 300–1000 kg would be too large for an armour layer of 3–6 t. An overtopping discharge of 5 l/s per m is reached for a crest freeboard of 3.45 m. Together with the design water level of +1 m CD, this gives a crest level of +4.5 m CD. This is even a little lower than the +4.8 m CD that is needed for a berm breakwater with 1–4 t rock, see Figure 8.13. A possible cross-section is shown in Figure 8.19.

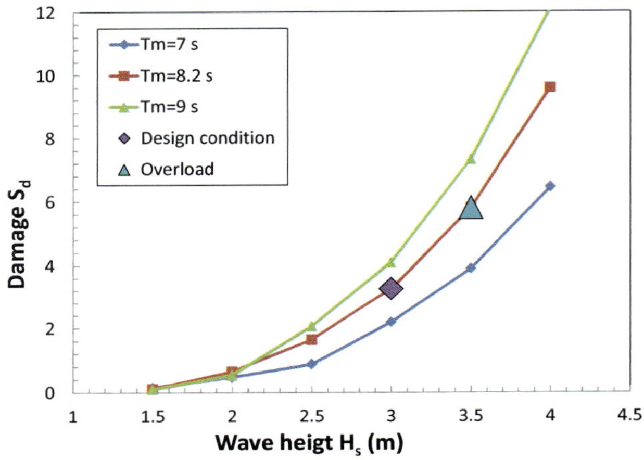

Figure 8.18. Damage curves for a conventional rock slope with an armour layer of 3–6 t; cotα = 1.75; P = 0.4; N = 3000.

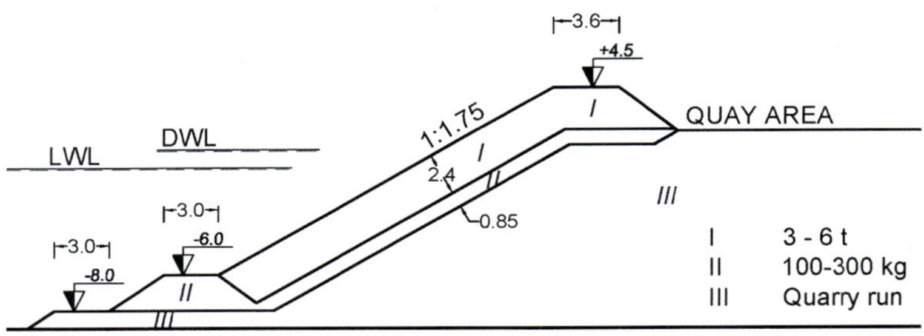

Figure 8.19. Conventional rock armoured structure with 3–6 t on a slope of 1:1.75. Design for H_{sD} = 3 m and q_{100y} = 5 l/s per m.

The design of the cross-section in Figure 8.19 can be compared with the berm breakwater design in Figure 8.14. (see Figure 8.20). The conventional design has a slightly larger armour rock, 3–6 t instead of Class I 1–4 t. The volume of large rock for the conventional design is still significantly larger than for the Class I rock: 63 m^3 per m length against 33 m^3 per m length. But the total volume of rock, taken seaward from the quay area, is larger for the berm breakwater (346 m^3 per m length) than for the conventional design (275 m^3 per m length). The slope of the conventional design is quite steep and the reach needed to place the lowest large rock is not too large. It can possibly be done by a large excavator.

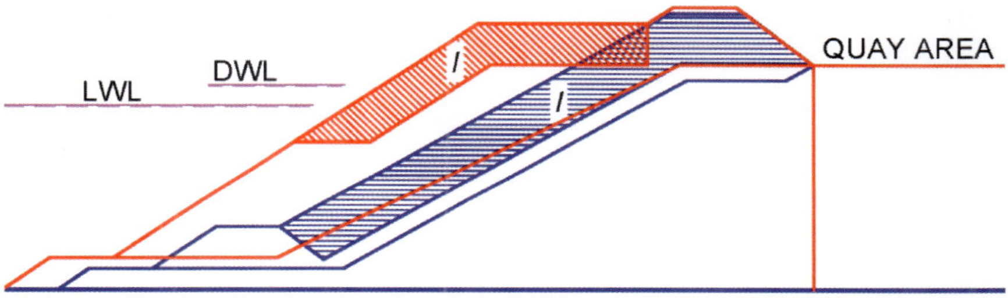

Figure 8.20. Comparison of conventional cross-section with a berm breakwater design.

Overall, the conventional design needs slightly larger rock, but the total volume of rock is substantially less than for the berm breakwater design. Both structures can easily be constructed.

8.3.4 *Overall conclusions and comparison*

A design wave height of H_{sD} = 3 m can be considered as a mild wave climate and a conventional design with a relatively steep slope of 1:1.75 and rock of 3–6 t is well able to resist such a wave climate. If a berm breakwater is designed, quite small rock is sufficient to make a proper design: a hardly reshaping berm breakwater comes to Class I of 1–4 t rock and a partly reshaping berm breakwater to only 0.5–2 t rock.

The usage of armour rock of a conventional design, compared to Class I of a berm breakwater, is always significantly larger (roughly a factor of 2). But the total volume of rock in this case of a mild wave climate (and relatively steep slope) is significantly less for the conventional design.

If rock of 3–6 t can be produced, then a conventional design as in Figure 8.19, may well be cheaper than a berm breakwater design like in

Figure 8.14 or Figure 8.16. Only if this kind of rock cannot be achieved, one can think of a berm breakwater with smaller rock.

This leads to the conclusion that if armour rock is readily available for a conventional design with a steep slope, then a berm breakwater may not be a cheaper solution, depending on the price differences between the different rock classes. A berm breakwater may be more attractive if a conventional design leads to too large rock (in large quantities) and with a gentle slope. Such a berm breakwater also needs large rock, but in a much smaller quantity (just Class I rock), and construction will be easier.

8.4 Examples for design a wave height of 7 m

8.4.1 *HR IC dedicated quarry, Class I 20–35 t*

The design condition is a quite severe design wave height of $H_{sD} = 7.0$ m with $s_{op} = 0.04$, giving $T_p = 10.6$ s. The overload condition is 15% higher than the design condition, giving $H_s = 8.0$ m at the design water level DWL. This overload percentage is a little lower than in previous sections and in real design may depend on the steepness of the curve of the extreme wave climate. Tides range between 0 m CD to + 2 m CD (2 m tidal range). The design water level (100-years condition) = surge + maximum tide = +4 m CD. Daily waves are quite moderate and a safety margin of $\Delta w = 1$ m above MHWS will be enough for construction. The foreshore is flat at -18 m CD. All input and calculated values, including design choices, are given in the spreadsheet in Appendix D.

The wanted resiliency is taken at 20% reshaping, which is the upper value for the range of hardly reshaping berm breakwaters. The allowable overtopping q = 10 l/s per m for the 100-years event and no restriction is given for the overload. This means that very significant overtopping is allowed with large overtopping volumes. One should design the crest and specifically the rear slope accordingly and physical model testing is a must in this case, to check the stability of the rear slope for overtopping waves. The crest level is only calculated for the 100-years condition and becomes 12.5 m CD. Finally, the mass density of seawater is $\rho_w = 1{,}030$ kg/m^3 and of rock $\rho_r = 2{,}700$ kg/m^3.

A Class I of 20–35 t is a very heavy rock grading, far beyond standard gradings. It needs a dedicated armourstone production with specific care in quarry yield production and blasting design. However, these kinds of rock gradings have been produced in the past. Guidance is given in Sections 6.1 and 6.4.

For a Class I of 20–35 t one can choose connecting classes as Class II = 10–20 t, Class III = 4–10 t and a Class IV = 1–4 t, all given in Table 8.10. The stability number is calculated as $H_s/\Delta D_{n50} = 1.99$ for a Class I of 20–35 t, which indeed gives a hardly reshaping berm breakwater, but very close to a partly reshaping one (the transition is at $H_s/\Delta D_{n50} = 2.0$). The cross-section as calculated by the spreadsheet is given in Figure 8.21.

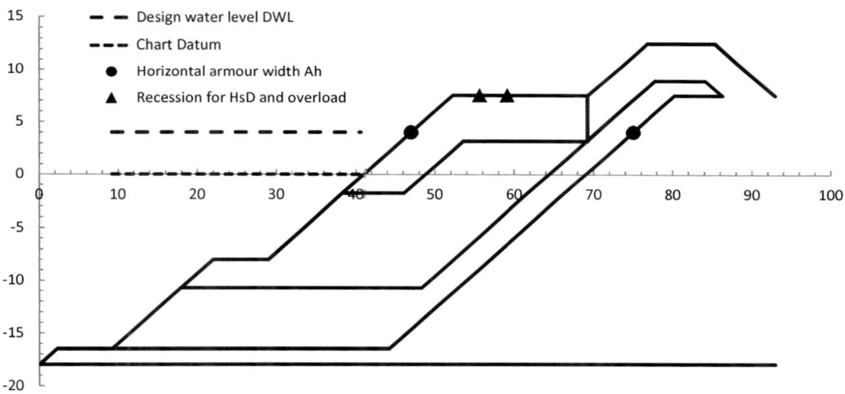

Summary of design choices	
Design of berm width	17.00 m
Design of berm level	7.50 m CD
Design of A_h	28.0 m
Design of transition class I to class I	-1.8 m CD
Design of crest level	12.50 m CD
Design of toe berm level	-8 m CD
Design choice cotα core below A_h	1.5 -

Rock Class	20	-	35	t
Rock Class	10	-	20	t
Rock Class	4	-	10	t

Figure 8.21. Calculated cross-section for H_{sD} = 7 m and Class I = 20–35 t.

The graph shows only three rock classes and not four (the spreadsheet gives only three classes). The geometry of Class I is as calculated, the other classes have to be distributed over the cross-section and this has been done in the final design in Figure 8.22. For a hardly reshaping berm breakwater, there is quite some expected recession of the berm, certainly for the overload condition (see Figure 8.21). This is due to the fact that the structure is close to a partly reshaping berm breakwater.

A toe berm has been designed at a level of -8 m CD, with Class III of 4–10 t as grading. For a hardly reshaping berm breakwater a toe berm is not necessary, but the structure is close to partly reshaping and displaced rock will fall onto the toe berm. Finally this will limit the amount of recession a little.

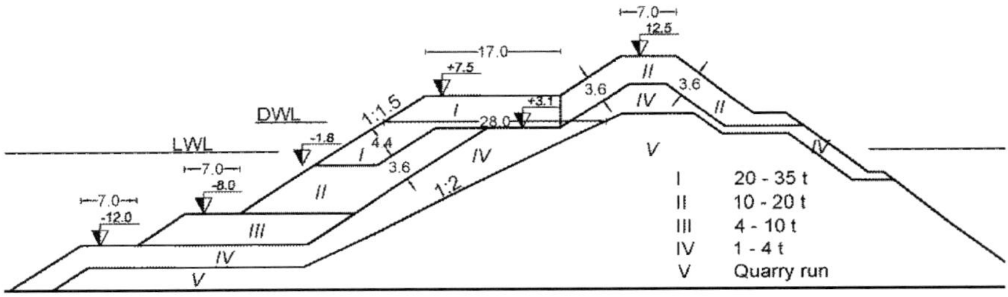

Figure 8.22. Hardly Reshaping Icelandic-type berm breakwater cross-section designed for $H_{sD} = 7.0$ m, Class I 20–35 t, $q_{100y} = 10$ l/s per m.

8.4.2 *PR IC dedicated quarry, Class I 10-20 t*

The design conditions are similar to the previous example. A short summary of these conditions is repeated here. All input and calculated values, including design choices, are given in the spreadsheet in Appendix D.

$H_{sD} = 7.0$ m $s_{op} = 0.04$ $T_p = 10.6$ s Overload $H_s = 8.0$ m
Tidal range 0 m CD to + 2 m CD DWL = +4 m CD
Safety $\Delta w = 1$ m above + 2 m CD Seabed at -18 m CD

The wanted resiliency is taken at 30% reshaping, which is the middle value for the range of partly reshaping berm breakwaters. The allowable overtopping is similar to the previous section: q = 10 l/s per m for the 100-years event and no restriction is given for the overload. The crest level is only calculated for the 100-years condition and becomes 12.5 m CD, similar to the previous example with Class I of 20–35 t. A Class I of 10–20 t is a heavy rock grading, beyond standard gradings. It needs a dedicated armourstone production with specific care in quarry yield production and blasting design. But these kinds of rock gradings have been produced in the past. Guidance is given in Sections 6.1 and 6.4.

For a Class I of 10–20 t one can choose connecting classes as Class II = 4–10 t and a Class III = 1–4 t, all given in Table 8.10. The stability number is calculated as $H_s/\Delta D_{n50} = 2.44$ for a Class I of 10–20 t, which indeed gives a partly reshaping berm breakwater, but quite close to a full reshaping one (the transition is at $H_s/\Delta D_{n50} = 2.5$). The cross-section as calculated by the spreadsheet is given in Figure 8.23.

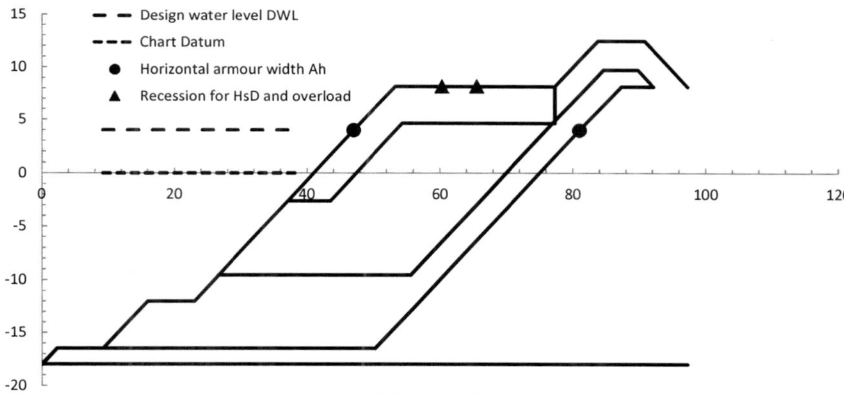

Summary of design choices	
Design of berm width	24.00 m
Design of berm level	8.20 m CD
Design of A_h	34.0 m
Design of transition class I to class I	-2.6 m CD
Design of crest level	12.50 m CD
Design of toe berm level	-12 m CD
Design choice cotα core below A_h	1.5 -

Rock Class	10	-	20	t
Rock Class	4	-	10	t
Rock Class	1	-	4	t

Figure 8.23. Calculated cross-section for H_{sD} = 7 m and Class I = 10–20 t.

The graph shows the three rock classes. The final design is shown in Figure 8.24. There is quite some expected recession of the berm, certainly for the overload condition (see Figure 8.23). It means that reshaping of the berm may well cut into the Class II rock underneath. In order to avoid increasing the thickness of the seaward side layer of 10–20 t, see the final design in Figure 8.24.

A toe berm has been designed at a level of -12 m CD, with Class III of 1–4 t as grading. This will limit the amount of recession a little. It is also possible to design a toe berm at -8 m CD, as in the previous example, but then the toe berm should be constructed of 4–10 t rock, which is Class II rock. In the final design the first choice has been made.

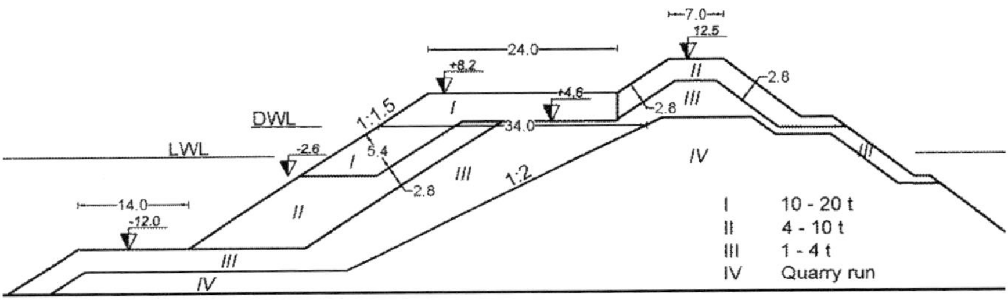

Figure 8.24. Partly Reshaping Icelandic-type berm breakwater cross-section designed for H_{sD} = 7.0 m, Class I 10–20 t, q_{100y} = 10 l/s per m.

8.4.3 *FR MA standard grading, Class I 6-10 t*

The design conditions are similar to the previous two examples. A short summary of these conditions is repeated here. All input and calculated values, including design choices, are given in the spreadsheet in Appendix D.

H_{sD} = 7.0 m	s_{op} = 0.04	T_p = 10.6 s	Overload H_s = 8.0 m
Tidal range 0 m CD to + 2 m CD			DWL = +4 m CD
Safety Δw = 1 m above + 2 m CD			Seabed at -18 m CD

The wanted resiliency is taken at 50% reshaping, which is 20% lower than the upper value for the range of fully reshaping berm breakwaters. The allowable overtopping is again q = 10 l/s per m for the 100-years event and no restriction is given for the overload. The crest level is only calculated for the 100-years condition and becomes 12.5 m CD, similar to the previous example with Class I of 10–20 t. As this is a fully reshaping mass-armoured berm breakwater, it is possible to use Equation 5.7 to check the stability of the rear side. For the overload condition $R_c/H_s * s_{op}^{1/3} = 0.36$, which still is well above the critical level of 0.21 that stands for start of damage. This means that a grading of 3–6 t over the crest and the rear side would be applicable.

In contrast to the two previous examples, the Class I grading of 6–10 t is a standard grading. The proposed mass-armoured berm breakwater has two classes and one can choose a connecting class as Class II = 3–6 t, both given in Table 8.10. If both gradings would be combined it results in a class 3–10 t, which is a kind of class that was used for the early designed fully reshaping mass-armoured berm breakwaters.

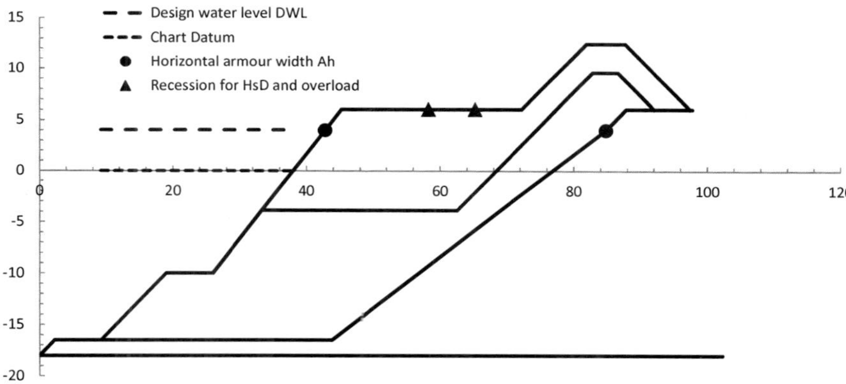

Summary of design choices	
Design of berm width	27.00 m
Design of berm level	6.00 m CD
Design of A_h	42.0 m
Design of transition class I to class I	-2.6 m CD
Design of crest level	12.50 m CD
Design of toe berm level	-10.00 m CD
Design choice cotα core below A_h	2.00 -

Rock Class	6 - 10 t
Rock Class	3 - 6 t

Figure 8.25. Calculated cross-section for H_{sD} = 7 m and Class I = 6–10 t.

The stability number is calculated as $H_s/\Delta D_{n50} = 3.01$ for a Class I of 6–10 t, which is really at the upper limit for fully reshaping mass-armoured berm breakwaters, as the transition is at $H_s/\Delta D_{n50} = 3.0$. The cross-section as calculated by the spreadsheet is given in Figure 8.25. As the stability number is very high, there is a lot of berm reshaping, with quite a big rock. One should use only good quality rock in this case.

A toe berm has been designed at a level of -10 m CD, with Class II of 3–6 t as grading. This will limit the amount of recession, see Section 3.7.4. The spreadsheet gives a standard toe berm width of one wave height (7 m), but given the expected recession of the berm it might be better to increase the width to about 10 m. In order to reduce the volume of 3–6 t rock it is also possible to use more core berm. This is the area drawn with dashed lines in the final design as given in Figure 8.26.

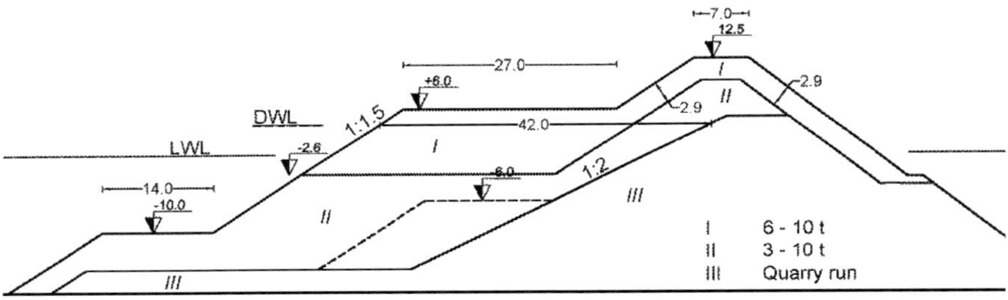

Figure 8.26. Fully Reshaping mass-armoured berm breakwater cross-section designed for $H_{sD} = 7.0$ m, Class I 6–10 t, $q_{100y} = 10$ l/s per m.

8.4.4 *Overall conclusions and comparison*

It is hardly possible to design a conventional rock armour layer for a wave climate with a design wave height of 7 m. From a stability point of view, it is only possible with the largest standard grading of 10–15 t in the Rock Manual (2007) and a gentle slope angle of 1:3.5. For a 10–20 t grading from a dedicated quarry, such a design still needs a slope angle of 1:3. For both examples it means that a crane is required with a reach of around 80–100 m that should place stones up to 20 t quite precisely in a double layer deep under water. Another option is to place the armour rock at the toe by barge, but that needs quite calm daily wave conditions to place the rock correctly.

Experience shows that conventional rock armoured structures with these large rock and with such gentle slopes have not been designed and constructed. The main reason is that in such a case one makes the choice to use concrete units on a much steeper slope.

But another option might of course be to design a berm breakwater, as was done in Sections 8.4.1 to 8.4.3. If a Class I of 20–35 t can be obtained from a dedicated quarry, this design (Section 8.4.1) is preferred as it gives the smallest total volume of rock and the largest resiliency. But a design with a Class I of 10–20 t (Section 8.4.2) also gives a good design. A fully reshaping mass-armoured design can be made of rock from a dedicated quarry, but also from standard gradings with Class I of 6–10 t, as was done in Section 8.4.3. In this case quite some recession will occur, which means that big rock is moving around from the berm downwards, with possible breakage of rock as a result. For this high design wave height of 7 m, a fully reshaping berm breakwater with only Class I rock of 6–10 t includes some (unknown) risks and if possible one should try to get a larger rock grading for the Class I rock.

"Rock" is the essential word in design of berm breakwaters. Often dedicated quarries can be found and opened to produce the required rock. This is different from designs with rock from existing quarries, where delivery of very large rock classes may be problematic. It has been proven possible to win really large rock in dedicated quarries and this experience has been described in Chapter 6. Quarry and project management as well as blasting and sorting techniques are essential in getting all required rock for an acceptable price. With this experience it is possible to design and construct berm breakwaters for a design wave height of 7 m or even more.

Construction of dedicated Icelandic-type of berm breakwaters is still fairly easy with excavators of maximum 120 t, which can handle rocks up to 35 t. Heavier excavators will probably become available on the market so the future may well be that rock over 35 t can be used in construction.

Chapter 9

Constructed Examples

9.1 Introduction

First designs of so-called modern berm breakwaters were made in 1983 and since then a significant number of berm breakwaters have been constructed all over the world, although with the majority being located in Iceland. The design guidance developed in this book was based on the experience gained from those earlier designs and constructions, as well as on the behaviour of these breakwaters over time.

This chapter describes a number of earlier designed and constructed structures and the sections have been divided in the hardly, partly and fully reshaping berm breakwaters.

9.2 Hardly reshaping berm breakwater

9.2.1 *The Dalvik berm breakwater, Iceland*

Dalvik is a fishing town in northern Iceland and its economy is mainly based on fishing and the fishing industry. In the early 1990s, the demand for better mooring conditions, both for the fishing fleet as well as for general cargo ships, called for more sheltered harbours. The wave agitation was studied in a 3D physical model on a scale of 1:60, which led to the planning of a 320 m long breakwater with a total volume of 104,000 m^3.

The town is located inside the fjord Eyjafjörður, which gives some shelter against the offshore wave climate. The design wave height at the breakwater was estimated as H_s= 2.5 m with a peak period of about

$T_p = 8$ s. The breakwater is located on a relatively flat sandy seabed with depths varying along the breakwater from -5 m to -9 m CD, with a foreshore slope in the wave direction of about 1:100. The tidal difference is about 1.4 m on spring tide.

A detailed search for an armourstone quarry was carried out. A potential quarry site was analysed and based on core drilling, a quarry yield prediction was established. The rock was of good quality, porphyritic basalt, with a saturated surface dry density of 2.9 t/m^3 and a point load index I_{s50} of 10 MPa (see Rock Manual [2007] for definitions). The yield prediction indicated that it would be possible to quarry up to 54% > 0.3 t, 24% > 1 t and 2% > 10 t.

With the moderate wave climate and relatively good quarry it was possible to use dredged material for the inner part of the core, coarse sand or gravel. This material was cheaper than to use quarry run and accounted for about 30% of the total volume of the structure. Ahead of the construction by land-based equipment, the dredged material was placed in a mound that reached up -2.0 m CD water depth, see the cross section in Figure 9.1. A picture of the construction is given in Figure 9.2. The top part of the mound consisted of gravel which could be placed with rather steep side slopes. This mound experienced a winter storm before being protected with quarried material from land, with only minor deformation measured. Photos in Figure 9.3 were taken during this storm event, showing waves breaking on the submerged mound in front of the land-based construction as well as on the partly constructed emerged breakwater.

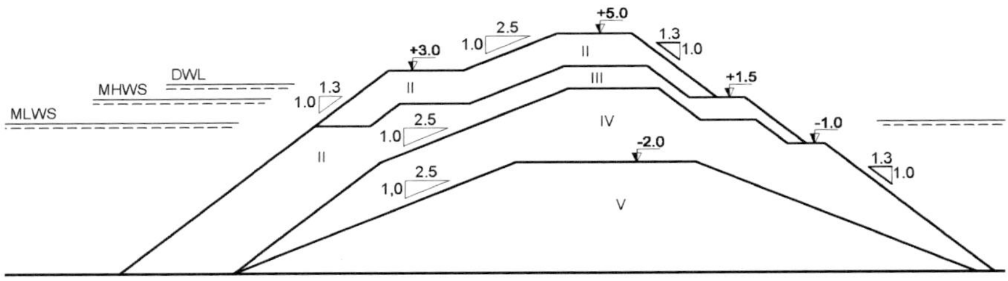

Figure 9.1. The design section of the trunk of the Dalvik breakwater from 1994.

Figure 9.2. Construction of the Dalvik breakwater during the winter 1994 to 1995.

Figure 9.3. The partly constructed Dalvik breakwater during a storm event in winter 1994 to 1995. Waves breaking on the emerged breakwater, left photo, and waves breaking on the submerged mound, right photo.

Of the remaining part of the cross-section, that is above the sand/gravel of the inner core, the outer core accounted for about 50% and the sorted rock for about 50%. As the predicted quarry yield larger than 0.3 t was about 50%, it was decided to utilise all stones larger than 0.3 t for the sorted rock part. The smallest stone class was chosen rather wide as Class III 0.3–1.5 t, to capture a high yield of about 30%. This class has a stability number $H_s/\Delta D_{n50}$ of 2.2. With such a low stability number the breakwater could have been designed with only one rock class. But the quarry was expected to yield about 20% larger than 1.5 t and it would be a waste not to utilise this for the benefit of the structure.

Therefore it was decided to reinforce the cross section of the trunk with stone Class II of 1.5–4.0 t rock with $H_s/\Delta D_{n50}$ = 1.5 (see Figure 9.1). Two layers from low water level on the sea side up to the crest were designed, with one layer on the rear side. Similarly the breakwater roundhead was strengthened with Class I of 4–8 t with a stability number of $H_s/\Delta D_{n50}$ = 1.1. According to the quarry yield prediction, this design fully utilised all stones from the quarry larger than 0.3 t, with an overproduction in both classes I and II. In spite of using up to 30% of dredged material for the breakwater, the design is achieving a 100% utilisation of all quarried material and delivering a very strong and robust breakwater.

The project was tendered out and 15 bids were received from 9 contractors with tender prices ranging from 60% to 85% of the client's cost estimate. The lowest bidder got the contract. The construction period extended over two summer periods, 1994 and 1995, but instead of halting the construction during the mid-winter the contractor chose to continue working through the whole winter. The new quarry was opened and with assistance from the design team, the contractor was able to fulfil the quarry yield prediction. The contractor delivered very good work with good craftsmanship in rock placing. After 20 years of service, not a single stone has moved on the breakwater (see Figure 9.4).

Figure 9.4. The Dalvik breakwater. Photo by Haraldur Gudjonson.

Table 9.1 gives design characteristics of the breakwater, together with values that come out by applying the geometrical design rules from Chapter 5. The Dalvik breakwater has a steep 1:1.3 lower slope with a quite gentle upper slope of 1:2.5. According to the geometrical design rules both upper and lower slope could be chosen as 1:1.5. Calculated recession is very small and the required berm width from geometrical rules could even be a little smaller. But a berm width of only 4 m is already quite a small berm.

The actual berm level was even a little lower than the calculated berm level (+3 m CD versus +3.8 m CD), meaning that the contractor took a little larger risk on damage during construction. The required crest level of +5.4 m CD, as calculated for an allowable overtopping discharge of 10 l/s per m, is very close to the realised crest level of +5.0 m CD.

A design based on the design rules and utilising the same stone classes as the actual design from 1994 is given in Figure 9.5. The actual design, given with fine grey lines, has a wider berm but slightly lower berm and crest levels. Only the exposed side has been changed and the rear side of the new design is left unchanged. To make a better comparison between the two designs, the intersection between stone Class III and the core below -2.0 m elevation is also unchanged and has a slope of 1:1.3.

In total the new cross section is 6% less voluminous compared to the actual design. But as for the stone classes they are 12% less voluminous. This indicates that the actual design from 1994 is on the safe side but could have been made less voluminous.

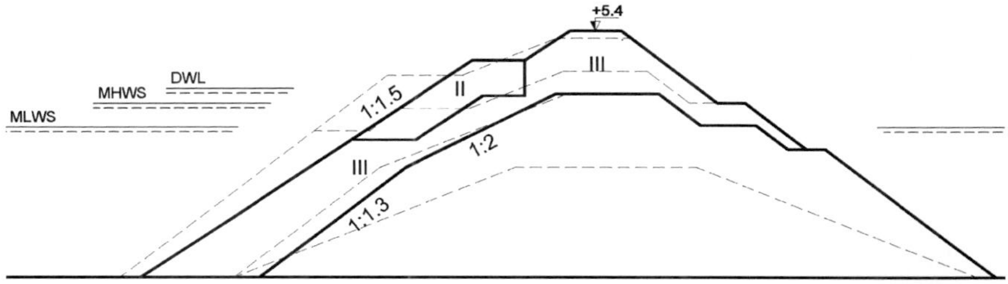

Figure 9.5. New cross section for the Dalvik breakwater based on the design rules and utilising the same stone classes as the actual design from 1994, which is shown with fine grey lines.

Table 9.1. Dalvík breakwater, trunk cross-section, comparison between the actual design and the geometrical design rules.

Design conditions	*Parameter*	*Actual design*	*Geometrical design rules*
Design wave height	H_{sD} (m)	2.5	2.5
Peak period	T_p (s)	8.0	8.0
Overload wave height	H_s (m)	3.0	3.0
Design water level, DWL	(m CD)	2.3	2.3
Allowable overtopping q for H_{sD}	(l/s per m)		10
Allowable overtopping q for H_{sOL}	(l/s per m)		50
Selected stone classes			
Class II, masst	(t)	1.5-4	1.5-4
Stability number of Class II, $H_s/\Delta D_{n50}$		1.5	1.5
Class III, weight	(t)	0.3–1.5	0.3–1.5
Stability number of Class III, $H_s/\Delta D_{n50}$		2.2	2.2
Type of berm breakwater		HR	HR
Front slopes			
Upper slope	cot α	2.5	1.5
Lower slope	cot α	1.3	1.5
Length parameters			
Recession for H_{sD}	Rec (m)		0.22
Recession for H_{sOL}	Rec (m)		0.48
Wanted resiliency	%		30
Min berm width from resiliency	B (m)		0.75
Min berm width from geometry	B (m)		2.8
Chosen berm width	B (m)	4.0	2.8
Min berm level from waves	d_b (m CD)		3.8
Min berm level from constructional issue	d_b (m CD)		3.8
Chosen berm level		3.0	3.8
Min horizontal armour width	A_h (m)	11.1	7.3
Min transition level between Classes I and II	$h_{I\text{-}II}$ (m)		-0.3
Crest level and crest width			
Overtopping influence factor	γ_{BB}		0.51
Required crest level for H_{sD}	R_c+DWL (m CD)		5.4
Required crest level for H_{sOL}	R_c+DWL (m CD)		5.1
Chosen crest level	(m CD)	5.0	5.4
Required crest width	(m)		

9.2.2 *The Helguvik berm breakwater, new design, Iceland*

In 2008–2009 the Helguvik berm breakwater, originally designed by Baird and Hall [1984] and constructed in 1986–88 (see Section 9.4.1), was extended by about a 170 m long spur close to the front end. With the closing of the Keflavik NATO naval base in 2006, the ownership of the Helguvik harbour had been transferred to the local harbour authority and the extension of the breakwater was a part of a plan to give shelter to a new bulk quay in connection with a planned industrial area. The angle between the original breakwater and the spur is about 107°, see Figure 9.6. Compared to the original breakwater with a design wave approach angle of about 45°, the approach angle for the breakwater extension is about 28° or closer to being perpendicular. This means that the wave load on the extension is probably higher.

Figure 9.6. The new hardly reshaping Icelandic-type berm breakwater at Helguvik, connected to the old partly reshaping mass-armoured berm breakwater. Photo by Haraldur Gudjonson.

With considerably more knowledge on structural behaviour of berm breakwaters, breakwater construction and utilisation of armourstone quarry, the new design is a much more compact design. The same armourstone quarry is used as for the original breakwater, but while the original design only demanded stones from 1.7 up to 7 t to be produced from the quarry, the new design utilised rock in the range from 0.3 to 25 t in 4 stone classes. The breakwater was constructed in 24 to 28 m water depth. Before the construction from land started, dredged material from rock dredging in front of the planned new quay was utilised to fill under the breakwater up to a level of -8 to -10 m. The design allowed the lower slopes to be as steep as 1:1.3 but it was anticipated that the slope would become gentler.

The excavators placing the armourstone were equipped with a GPS system coupled with a 3D model of all layers of the breakwater and displayed in front of the operator. This enabled the operator to achieve even placement of armourstone in accordance with the design lines, (see Figure 9.7).

Figure 9.7. The new hardly reshaping Icelandic-type berm breakwater at Helguvik.

9.3 Partly reshaping berm breakwater—Icelandic-type

9.3.1 *The Sirevåg berm breakwater, Norway*

The Sirevåg breakwater is located in a narrow bay on the west coast of southern Norway (see Figure 9.8) and was constructed in the period from January 2000 to July 2001 [Sigurdarson *et al.*, 2003]. The breakwater was designed by the Icelandic Maritime Administration for the

Norwegian Coastal Directorate. It was designed as a partly reshaping Icelandic-type berm breakwater for a wave height with a 100-years return period. The established design criteria were that the recession of the trunk should be within $2D_{n50}$ and, on the roundhead, within $3D_{n50}$. The breakwater should also withstand a wave height with a 1000-years return period, which was referred to as the worst-case scenario, without total failure.

The design 100-years recurrence wave height at the location of the breakwater was established by SINTEF as H_{sD} = 7.0 m with T_p = 14.2 s and the worst case scenario as H_s = 7.7 m with T_p = 15 s.

Figure 9.8. The Sirevåg breakwater shortly after construction. The main quarry areas can be seen on both sides of the bay, the larger on the opposite side. Courtesy of the Norwegian Coastal Administration.

The breakwater has partly been located on a rocky bottom and partly on fine quartz sand. The depth of the rocky bottom is very variable from 3 m to 22 m, with steep slopes. Under the outermost 150 m a flat sand bottom at -18 m CD is present. With a very small tidal difference, mean high water spring tide at +0.2 m CD, the design high water level associated with the 100-years design storm was established at +1.3 m. The breakwater is about 500 m long with a total volume of approximately 640,000 m^3.

Three main geometrical design parameters were considered, the armour width, A_h, the berm level, d_b, and the crest level, R_c. In the design the dimensionless armour width parameter, A_h/H_{sD}, was chosen as 4.5 for the most exposed trunk section but 5.0 for the head section. This is on the safe side as with present knowledge, as presented in Equation 5.9, it could have been taken as 4.2. The berm level assumed a working level for excavators at +1.4 m CD. With two layers of Class I rock on top of this for the trunk section and an additional Class III layer on the head section, the dimensionless berm level above design water level resulted in $d_b/H_{sD} = 0.6$ for the most exposed trunk section and 0.79 for the head section. The design practice at that time for the dimensionless crest level, with no activity on the lee side of the breakwater crest, was $R_c/H_s = 1.0$ to 1.2. For the design of the Sirevåg breakwater it was decided to raise this parameter to 1.25. It should be realised that as the berm width, B, was not a critical design parameter, the crest height and width has only limited influence on the cross-sectional volume and therefore cost, as it only affects the crest volume above berm level. A higher crest level or a wider crest would result in a narrower berm, while the armour width, A_h, remains constant.

In the preliminary design three sets of stone classes were considered based on initial size distribution estimates from the potential quarries. In the final design, after thorough quarry investigations, one set was chosen based on the overall utilisation of all quarried material according to a quarry yield prediction and fulfillment of stability criteria for all sections of the breakwater (see Table 9.2).

Class I rock was chosen to be 20–30 t, with $M_{50} = 23.3$ t according to the definition in Section 6.2.4 and Equation 6.4. This is a safe approach for the M_{50} and led to $H_{sD}/\Delta D_{n50} = 2.1$ for design conditions, which is just in the class of partly reshaping berm breakwaters (see Section 2.3).

Table 9.2. The Sirevåg breakwater stone classes and quarry yield prediction.

Stone class	M_{min}-M_{max} (t)	M_{50} (t)	M_{max}/M_{min}	D_{nmax}/D_{nmin}	Expected quarry yield
I	20–30	23.3	1.5	1.14	5.6%
II	10–20	13.3	2.0	1.26	9.9%
III	4 – 10	6.0	2.5	1.36	13.7%
IV	1 – 4	2.0	4.0	1.59	19.3%

The final design consisted of 10 different cross sections, where Class I is the largest stone class on six of those, Class II on two and Class III the largest stone class on two sections. A cross-section of the most exposed trunk section is shown in Figure 9.9. The design fully utilised all quarried stones over 1 t and a 100% utilisation of all quarried material was expected for the project. The Class I stones were specified to be orderly placed with interlocking.

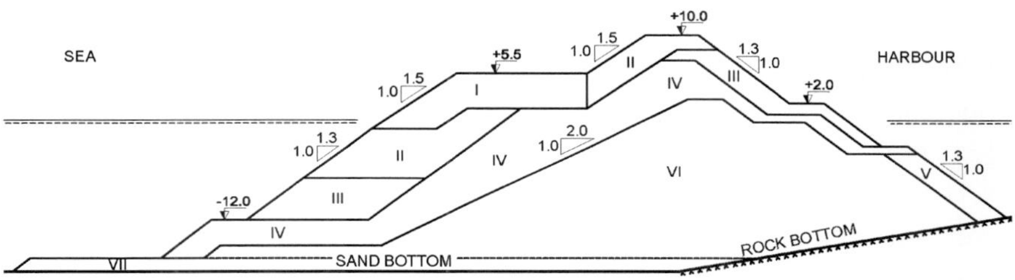

Figure 9.9. Sirevåg berm breakwater, cross section of the outer part.

The design was based on earlier design of berm breakwaters in Iceland by the design team. As such it had background in stability tests performed for some of those projects, but the Norwegian Coastal Directorate did not deem it necessary to verify the design by model tests before the project was tendered out. After the construction had started, model tests were undertaken at SINTEF/NTNU as a diploma thesis study of students [Tørum *et al.*, 2003].

The model tests were carried out in a 5 m wide and 40 m long wave flume on a fairly small scale of 1:70. The modelled breakwater followed the drawings and design specification of the breakwater, then under construction, with one exception. The stones on the top of and at the front of the berm above elevation -1.0 m CD were placed randomly or pell-mell in the model, instead of an orderly placement. The tests confirmed the stability of the breakwater and the results were used to compare with the prototype experience, but they were not a part of the design process. According to these tests the dimensionless recession, Rec/H_{sD}, for the 100-years design condition was between 1.5 and 2, and for the 1000-years conditions between 2 and 2.5.

The Sirevåg breakwater was hit by a severe storm on January 28, 2002, only six months after it was finished. A Waverider buoy, located 450 m off the breakwater head at 20 m water depth, measured wave heights at half hour intervals. The maximum recorded significant wave height was H_s = 9.3 m and the wave height exceeded H_s = 8.0 m for a period of 3 hours. Sigurdarson *et al.*, [2005-b] have estimated the wave height at the breakwater to be about 88% of the wave height at the buoy and if compensated by wave reflection from nearby coastal areas the maximum wave height at the breakwater is estimated H_s = 7.9 m and exceeded H_s = 6.8 m for a period of 3 hours, which is close to the 100-year design wave conditions. Tørum *et al.*, [2005] estimated the wave height at the breakwater in the range of H_s = 7.1–8.7 m, which exceeded the 100-years design wave height. With both these estimates taken into account, the expected dimensionless recession should then have been close to 2.

Inspection of the breakwater shortly after the January 2002 storm showed that in three areas stones at the still water line were displaced, one area at the roundhead and two on the trunk. At the time of the inspection the reshaping had yet not progressed upwards to the top of the berm and there was no measureable recession.

On 11 to 12 January 2005 the breakwater was hit by major storm "Inga". At this time there were no wave measurements outside the Sirevåg breakwater. Tørum *et al.*, [2005] estimated the wave height at the breakwater to be in the range H_s = 6.3–7.7m. These estimates were based on wave measurements about 60 km north of Sirevåg, calibrated wave hindcast closer to the breakwater and wave refraction up to the breakwater. Eyewitness accounts from experienced seamen in Sirevåg stated that the waves impacting the breakwater during "Inga" were heavier and more powerful compared to the 2002 storm. During "Inga" damages occurred on old breakwaters, seawalls and lighthouses, up and down the coast to Sirevåg, the oldest structures from the early 20th century. The design team regarded this storm as being close to the upper limit of the estimate of Tørum *et al.*, [2005] or close to H_s = 7.7 m, which is the 1000-years wave condition.

This means that during these four years the breakwater was twice exposed to waves close to or exceeding the 100-years wave height and even close to the 1000-years wave height. This suggests an under-prediction of the design wave height.

During a site visit to the Sirevåg breakwater in September 2008, the breakwater was inspected. The outer part of the breakwater and the breakwater head have suffered recession, which was more extensive than after the 2002 storm. An estimate of the recession is presented in Figure 9.10, where a recent aerial photo has been overlain on the original design drawing. On the breakwater trunk the maximum recession is of the order 2.1 m to 6.2 m. With a mean diameter of Class I armourstones of 2.05 m, this corresponds to a recession of about 1 to $3D_{n50}$. On the breakwater head the maximum recession is about 8.4 m, which corresponds to $4D_{n50}$.

After the first storm, the start of reshaping near the still water level had not progressed up to the top of the berm, the reshaping after the two storms is very close to what could be expected from the model tests. There is an overall good agreement between the prototype experience and the model testing. With the two storms being close to or exceeding the 100-years design conditions, it can also be concluded that the design criteria are met.

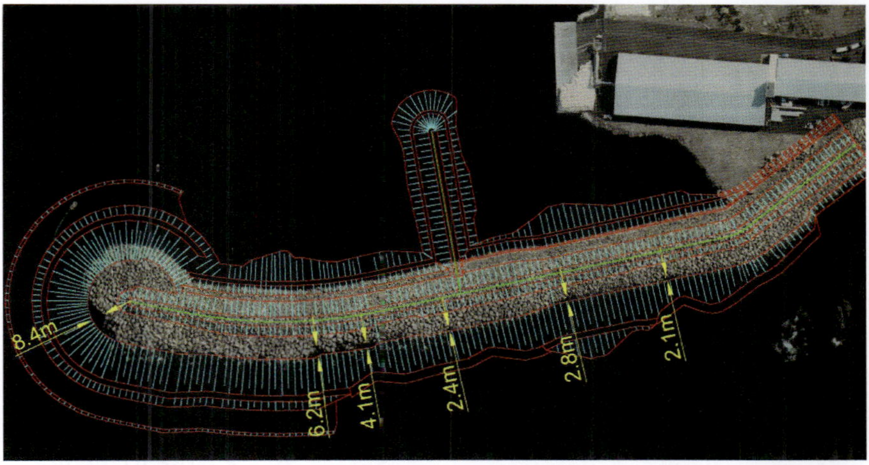

Figure 9.10. Sirevåg breakwater, estimated recession from an aerial photo.

9.3.2 *The Hammerfest berm breakwater, Norway*

An Icelandic-type berm breakwater was chosen to protect the Hammerfest LNG plant in northern Norway, located on a small island Melkøya (Figure 9.11). About 2,300,000 m^3 of solid rock had to be blasted to level the island and to construct a 900 m long berm breakwater or seawall to protect the plant.

Figure 9.11. The Hammerfest LNG plant in November 2003. The breakwater revetment extends from the rocky shore in the centre left of the island to the breakwater head in the lower right hand corner and protects the plant area as well as the harbour. The most exposed part is leftward facing revetment. Courtesy of Statoil.

Quarry investigations based on core drillings instead of a test blasting, were used to estimate the possible yield from the quarrying. It was thought to be possible to obtain 3% to 5% of a stone class of 20 t to 35 t and 26% to 35% larger than 0.5 t. About 1,500,000 m^3 of solid rock was required for the armourstone production (see Figure 9.12). The contractor developed blasting designs with the goal of achieving fragmentation as close to the design curve as possible. The rock excavation, breakwater construction and levelling of the island was carried out in a nine-month period from July 2002 to April 2003, except for the landfall of the gas pipes, which was finished during the summer of 2003. The maximum weekly production exceeded 100,000 m^3. The major challenges in the execution phase were the very tight construction schedule, producing the required stone classes, stockpiling in a very limited area and simultaneous construction of the breakwater.

Figure 9.12. In the Hammerfest project banded gneiss rock was used to produce armourstone up to 35 t. Thousands of years of weathering has emphasized weaknesses in the rock due to variations in mineral concentrations.

The breakwater was designed as a partly reshaping berm structure for the 100-years storm event of H_s = 7.5 m and T_p = 15.6 s. It should also withstand a storm with 1000-years return period of H_s = 8.5 m and T_p = 17.0 s, without total reshaping of the berm or breakwater. The final design consisted of 5 stone classes with Class I 20–35 t (see Figure 9.13 and Table 9.3).

For the 100-years design wave height the stability number of stone Class I is $H_s/\Delta D_{n50}$ = 2.2, which means that the structure is partly reshaping according to the classification in Section 2.3. If the structure would have been designed as a homogeneous mass-armoured berm instead of a multilayered Icelandic-type berm, it would become a fully reshaping structure with stability number $H_s/\Delta D_{n50} = 3.0$. In that case Classes I–V were taken together (0.5–35 t) with a mean weight of 9.7 t.

Table 9.3. The Hammerfest breakwater. Stone classes and quarry yield prediction.

Stone class	M_{min}-M_{max} *(t)*	M_{50} *(t)*	M_{max}/M_{min}	D_{nmax}/D_{nmin}	*Expected quarry yield*
I	20–35	25.0	1.75	1.21	3.5%
II	10–20	13.3	2.0	1.26	4.5%
III	4–10	6.0	2.5	1.36	6.5%
IV	1.5–4	2.3	2.7	1.39	7.5%
V	0.5–1.5	0.8	3.0	1.44	9.0%

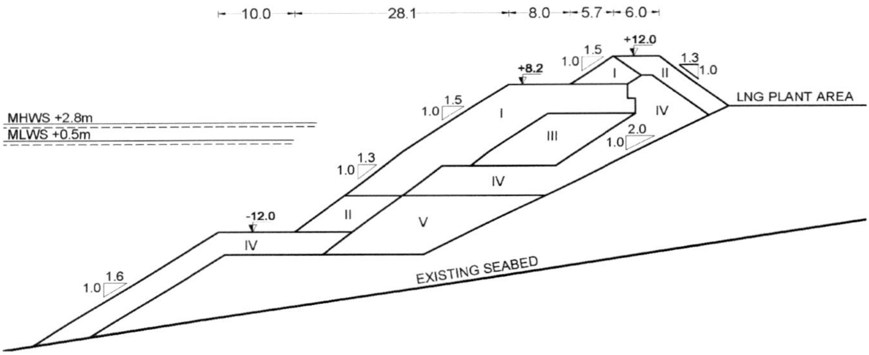

Figure 9.13. Cross-section of the most exposed part of the breakwater protecting the Hammerfest LNG plant.

During an inspection in the autumn of 2011, eight years after the structure was finished, it had not been hit by any major storm and no damage or displacements of armourstone had been recorded. Figure 9.14 shows a picture of the breakwater. The porous high berm protects the potentially vulnerable plant against overtopping and sea spray icing in arctic environment.

Figure 9.14. The Icelandic-type berm structure protecting the Hammerfest LNG plant, Norway. Photo taken in 2008.

9.3.3 *The Husavik berm breakwater, Iceland*

Husavik is a fishing village and commercial centre for the agricultural areas in Northeast Iceland. Recently Husavik has become known as the capital of whale watching in Iceland. As Husavik is located on the eastern side of the Skjalfandi Bay, the harbour is open to waves approaching from the northwest while the offshore wave conditions show increasing wave height from northwest to northeast. Two berm breakwaters have been constructed in Husavik to shelter the harbour.

In 1988 and 1990 an existing pier, consisting of a 10 m wide old concrete caisson, was protected with a berm breakwater on the sea side, [Sigurdarson *et al.*, 1994], see the structure in the background of Figure 9.15. This was partly to protect the pier structure, partly to provide more space for cargo handling and partly to diminish wave overtopping that often caused problems for cargo handling. Construction of the berm breakwater was spread over a period of two years, due to constraints in funding, with the partly protected structure having to survive over two winter periods.

The design wave height for this structure was established as $H_s = 4.0$ m with a peak period of $T_p = 15.5$ s and a design water level of +2.7 m CD. The design of the berm was of an early type mass-armoured berm breakwater, consisting of one stone class of 1–5 t, a low berm and rather steep front face. With the resulting stability parameter of $H_s/\Delta D_{n50} = 2.6$ the structure is close to transition between partly and fully reshaping mass-armoured berm breakwater. Although it experienced a heavy storm in December 1992, which was regarded as the worst for decades, the structure did not experience any reshaping during the 11 years it was the main shelter for the harbour.

Figure 9.15. Husavik harbour in 2009. In the centre the 1988–1990 breakwater protects an older concrete pier. In the foreground the 2001–2002 breakwater. Photo by Mats Wibe Lund.

However, the 1988–1990 berm breakwater did not extend around the existing pier and as such did not improve the wave agitation in the harbour. As the harbour entrance was rather wide, wave agitation including large long period oscillations and ship movements in the harbour often exceeded the acceptable criteria. The need to reduce agitation in the harbour as well as the need for a new quay for larger ships serving a planned industrial area, called for the construction of a new breakwater. Several proposals were studied in a 3D physical model study. These included lengthening of the existing breakwater, which limited the size of ships capable of entering the harbour, or a new more exposed breakwater. The chosen layout consisted of a new 350 m-long outer breakwater with a 130 m-long quay on its harbour side planned for a 10 m water depth.

The more exposed Husavik breakwater was built in 2001–2002, [Sigurdarson *et al.*, 2006], with a design wave height of $H_s = 6.8$ m and $T_p = 15.5$ s. The armourstone quarry that had been used for earlier rubble mound structures in Husavik, located only 5 km from the harbour, was not able to yield armourstone suitable for the increased wave height. Therefore, a new armourstone quarry was opened 25 km from the construction site, where all armourstone heavier than 1 t were quarried. The rock type was basalt of good quality with a high density of 2900 kg/m^3. Smaller armourstone and quarry run came from the old quarry.

The largest stone class chosen for the design was 16–30 t with a median weight of 20.7 t, corresponding to a stability parameter of $H_s/\Delta D_{n50} = 1.9$. See Figure 9.16 for placement of these large armourstone. To get the best utilisation of the quarried material, it was decided to use 5 stone classes for the breakwater (see Table 9.4). The total volume of the breakwater was roughly 275,000 m^3, about 140,000 m^3 armourstone and 135,000 m^3 quarry run.

Figure 9.16. Excavator placing Class I 16–30 t stones on the head of the Husavik berm breakwater, Iceland.

Table 9.4. Breakwater stone classes and quarry yield prediction for the Husavik breakwater, constructed in 2001–2002.

Stone class	M_{min}-M_{max} (t)	M_{50} (t)	M_{max}/ M_{min}	D_{nmax}/ D_{nmin}	Expected quarry yield
I	16.0–30.0	20.7	1.9	1.23	5%
II	10.0–16.0	12.0	1.6	1.17	5%
III	4.0 – 10.0	6.0	2.5	1.36	9%
IV	1.0 – 4.0	2.0	4.0	1.59	14%
V	0.3 – 1.0	0.5	3.3	1.49	12%

Figure 9.17. The partly reshaping Icelandic-type berm breakwater at Husavik 14 years after construction. Based on 3D physical model study, the shallows at the root of the breakwater cause wave breaking and high wave forces that were avoided with the breakwater alignment.

On 25 October 2008 the breakwater was hit by a storm with an offshore significant wave height exceeding 11 m and with a peak period of about 13.5 s, [Sigurdarson and van der Meer, 2012]. There was flooding in the harbour area and minor damages on rock revetments inside the port. According to a wave hindcast study the wave height reaching the breakwater was about H_s = 5.0 m, corresponding to a stability number $H_s/\Delta D_{n50} \approx 1.5$ for the Class I rock. There were no damages on the breakwater itself, not a single stone displaced. Still the breakwater was heavily overtopped and gravel and small stones from an unpaved area behind the crest were washed onto the paved quay area.

Figure 9.18. The Husavik breakwater during a storm in October 2014, waves breaking on the slope and overtopping. Photo by Gaukur Hjartarson.

Figure 9.19. Waves breaking in front of the breakwater roundhead and overtopping the Husavik breakwater in October 2014. Photo by Gaukur Hjartarson.

In October 2014 the breakwater was again exposed to heavy waves breaking on the berm and overtopping (see Figure 9.18 and Figure 9.19). As for the 2008 storm, not a single stone was dislocated.

In 2008 feasibility studies for a new aluminium smelter near Husavik were undertaken. These included a new berth capable to accommodate a Panamax bulk carrier. Several locations were looked at and the chosen option included a still more exposed breakwater with a design wave height of $H_s = 8.0$ m, [Sigurdarson *et al.*, 2008]. These studies included a comparison between a berm breakwater reinforced with a 20 to 50 t stone class and a 29 t Xbloc interlocking concrete block. This project has, however, not been realised.

9.4 Partly reshaping berm breakwater - mass-armoured

9.4.1 *The Helguvik berm breakwater, original design, Iceland*

The berm concept was introduced in Iceland through the design phase of the 900,000 m^3 Helguvik breakwater in the early 1980s, [Baird and Hall, 1984] and [Baird and Woodrow, [1987]. The breakwater shelters the Helguvik Bay for a tanker terminal close to the former Keflavik NATO naval base. It was built for the U.S. Navy with a design contract with Bernard Johnson Inc. which retained W.F. Baird and Associates to assist in the development of the design. This was early in the development phase of berm breakwaters and it was a breakthrough for the concept when the design of the Helguvik breakwater was accepted by the military authorities. Physical model studies were undertaken by NRC in Ottawa and were overseen by the Icelandic Harbour Authority. The breakwater was constructed in 1986–1988 by a joint venture of Icelandic contractors.

The wave conditions in Helguvik Bay consist of locally generated northeasterly wind waves in the Faxafloi Bay, while the Reykjanes Peninsula shelters for the more heavy westerly and southwesterly offshore waves. The 50-years design wave conditions were established as a significant wave height $H_s = 5.8$ m with a peak period of $T_p = 9.6$ s with waves attacking the breakwater at approximately 45° angle. With mean high water spring tide of +4.0 m CD, the design water level was established at +5.0 m CD. The established design wave conditions have been regarded as rather conservative. Recently, nearby berm structures, as well as the extension of the Helguvik breakwater, (see Section 9.2.2) constructed by the local harbour authority have been based on lower design wave height.

As the early design of berm breakwaters, the design can be characterised by a low and wide berm of only one stone class of 1.7 to 7.0 t with a mean mass of 3.2 to 4.2 t. The design assumed that the breakwater could be built with a steep front slope. In case of natural slope being steeper than 1:1.5 the width of the berm should be increased to an extent that the volume lying above the slope 1:1.5 should be equal to the volume under the same slope not filled, relative to the theoretical berm width of 14 m. This resulted in a horizontal armour width on design water level, A_h, of about 23 m, assuming a 1:1.5 front slope. The cross-section of the original design is given in Figure 9.20.

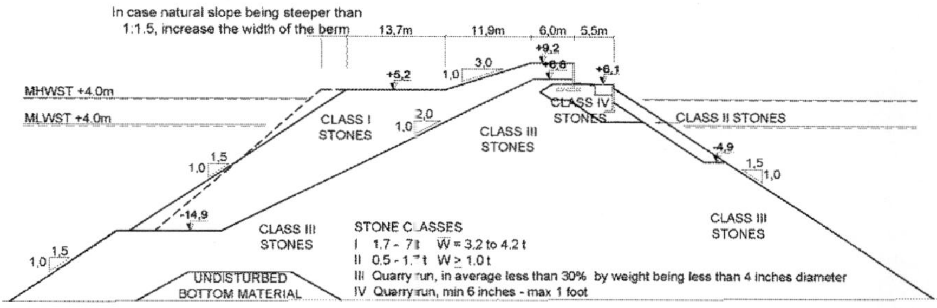

Figure 9.20. A cross-section of the Helguvik breakwater.

For the assumed design wave height, the stability number $H_s/\Delta D_{n50}$ of the stone class is probably close to 3.0, based on an assumed mean mass of 3.8 t and a mass density of 2800 kg/m^3 for the basaltic rock.

All material for the breakwater was quarried in a dedicated rock quarry opened adjacent to the breakwater site. Although the design was developed for the location using the expected quarry yield, at the time of construction, there was limited knowledge of quarrying for large armourstone. Still the quarry delivered considerable quantities of stones larger than what was needed for the design.

After nearly 30 years in service the breakwater has probably not been exposed to higher waves than about H_s = 4.0 m. With the same assumed mean weight of armourstone as above, the breakwater has not experienced higher stability numbers higher than about 2.1. For this wave height the horizontal armour width, A_h, would only need to be about 17 m, compared to the 23 m of the actual design. The breakwater has performed well with very limited reshaping and only few stones from the front slope relocated, probably due to the initially steep front slope. Although the intention was to have a reshaping berm breakwater, the berm has hardly reshaped at all during this period and probably quite some safety was applied to the detailed design as it was for the US Navy. A picture of the breakwater, taken in 2008, is given in Figure 9.21.

Figure 9.21. The partly reshaping mass-armoured berm breakwater at Helguvik, constructed in from 1986–1988. Picture taken in 2008.

In 1996, a berm structure was built to protect an old caisson breakwater in Keflavik, just 2 km from Helguvik. The wave load at the site is nearly the same as in Helguvik and the same quarry was used. The structure was designed according to the practise for Icelandic berm breakwaters with several stone classes. While the Helguvik breakwater only used rock size from 1.7 to 7.0 t, the Keflavik breakwater used all rock from 0.2 to 12 t, which means that the utilisation of quarried material was much better. Still the Icelandic design of the Keflavik breakwater used only about 40% of the total rock mass compared to the Helguvik breakwater, per cross-sectional area.

9.5 Fully reshaping berm breakwater

9.5.1 *The St George breakwaters, Alaska*

This section was written with the help of Jena Gilman, COWI Marine North America.

The St George berm breakwaters in the Pribilof Island Chain of Alaska's Bering Sea are examples of reshaping berm breakwaters [Gilman, 1987, 1999]. The breakwaters were originally designed by Peratrovich Nottingham & Drage, Inc (PND), in the early 1980s. The harbour provides moorage and safe refuge for the local fishing fleet (see Figure 9.22).

Figure 9.22. The St. George berm breakwaters; an aerial photo from the year 2000. Photo by J. Gilman.

Physical conditions at St. George are extremely harsh, with heavy icing and offshore waves that reach up to 15 m high. It means that large waves reach the structure almost every week during the winter time.

Initial testing of the fully reshaping berm breakwater design was performed at DHI, Denmark, see Juhl and Jensen [1995], who actually describe eight fully reshaping berm breakwater structures. At DHI the tested section consisted of a 16.8 m wide berm with rock 1.5-9 t and an average mass of M_{50} = 4.8 t. This gives a stability number of $H_s/\Delta D_{n50} = 3.3$, just above the limit for a statically stable fully reshaping berm breakwater.

After the failure of a berm breakwater design during construction at the sister island St Paul, the design was tested again in 1985, now at large scale in Delft Hydraulics' (now Deltares') large wave flume at a scale of 1:7 and on a smaller scale at 1:35. The design was finalised with 3D-testing in a wave basin. Some of the results have been published by Van der Meer [1992] and Van der Meer and Veldman [1992]. Details of results of the model testing have been given in Figure 4.20 (wave transmission over the structure in the Delta flume), Figure 5.5 (resulting profiles of the berm recession), Figure 5.6 (testing the berm recession in the Delta flume), Figure 5.12 (subsidence of rock if no apron is present) and Figure 5.17 (berm recession at the roundhead). The design tested at Delft Hydraulics also had a 16.8 m wide berm and the stability number for the depth-limited overload condition was $H_s/\Delta D_{n50} = 3.1$ and for the design condition just a little lower.

The horizontal armour width on design water level of the St George breakwater, A_h, is about 30 m. According to Equation 5.9, the required width would have to be about 37 m, assuming a design wave height of $H_s = 6.0$m.

The berm breakwaters were built during the period 1985–1987 of a locally quarried basaltic rock with a mass density of about 2700 kg/m^3. One of the first aerial photos of the breakwaters after construction is given in Figure 9.23. The picture is taken from more or less the same position as for the picture in Figure 9.22 and shows now the heavy wave breaking all over the foreshore in front of the breakwater. Heavily overtopping waves are shown in Figure 9.24.

Figure 9.23. Aerial view of the St. George breakwaters when near completion during a severe storm evaluated as equivalent of the design storm at the time. Photo and information by J. Gilman.

Figure 9.24. Waves overtopping the berm breakwaters at St. George.

Although these breakwaters have frequently been exposed to waves close to the design wave conditions (depth-limited condition), they have been functioning well over a long period of time. During the first years, individual stones did move on the slope but then became more locked. The advanced profile development during the first fifteen years, according to Gilman [1999], has been due to gradual settling of the entire mass as the toe has slowly eroded over the years. Inspection in 2001 showed that the profile had reshaped significantly and the berm on the South Breakwater would need to be restored in the next years after 2001.

After 19 years of storms, a massive storm hit the breakwaters and caused non-critical breakwater damage in October 2004. On the southern breakwater arm or trunk, from the elbow up to the roundhead, the full berm width was eroded and the recession had reached the upper slope and even the crest at some locations (see Figure 9.25). Wear and tear of the armourstone on the slope is evident and many stones on the slope are rounded, indicating frequent movement.

Figure 9.25. The St. George southern breakwater before the repair in 2006. The full berm width has eroded and the recession has reached the upper slope and even the crest at some locations. Wear and tear off the armourstone on the slope is evident. Courtesy of PND Engineers.

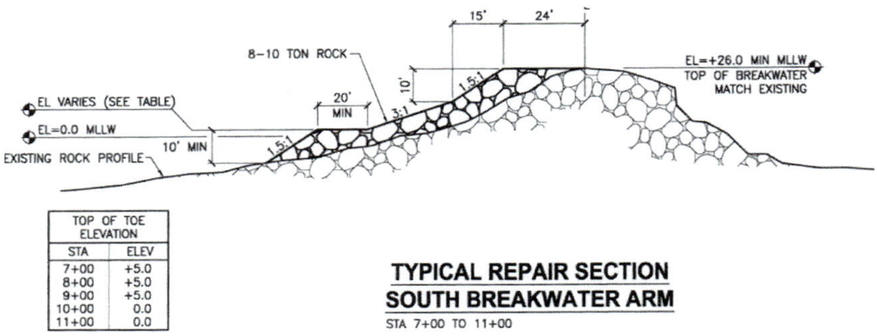

TOP OF TOE ELEVATION	
STA	ELEV
7+00	+5.0
8+00	+5.0
9+00	+5.0
10+00	0.0
11+00	0.0

Figure 9.26. Design of a typical repair section of the St. George southern breakwater in 2006, where 8–10 t rock was added to the profile. Courtesy of PND Engineers.

Due to the exceptional past performance, PND was asked to provide design repairs. A design cross section is shown in Figure 9.26, where the profile is strengthened with 8–10 t rock and the repaired profile is shown on Figure 9.27 (information from PND). Compared to the original rock, the average mass used for the repair is nearly twice as heavy. It is noteworthy how closely the repair design resembles the repair of the Bakkafjordur berm breakwater 10 years earlier, as seen in the next section and Figure 9.39. Although it is of course a classical two-diameter thick rock slope design, it is now on a reshaped and almost stable S-shaped profile.

Figure 9.27. The St George southern breakwater after the repair in 2006, where freshly quarried 8–10 t rock have been added to the profile. Courtesy of PND Engineers.

Although the design of this fully reshaping berm breakwater was thoroughly tested, it needed repair after about 20 years of service life. One main reason may be that the location at St George is frequently attacked by high waves during winter time and the design condition being probably closely reached once per year at a minimum. These were not the offshore design conditions, but as the situation is depth-limited, the conditions at the structure were not far from it. Testing in the laboratory was not performed by generating hundreds of storms just below design conditions, but by increasing wave conditions per sub-test of six hours. After such a test sequence, including the overload condition, the berm was almost fully reshaped, not leaving space for much more recession.

It is for this reason that the choice of a berm width for a fully reshaping berm breakwater in Section 5.2.2 was based on a recession up to design conditions, that would not be more than 70% of the total berm width, see Equation 5.3. For the fully reshaping berm breakwaters described in the next sections, repair was needed. It clearly shows the smaller resiliency of fully reshaping berm breakwaters versus the hardly or partly reshaping berm breakwaters.

9.5.2 *The Bakkafjordur breakwater, Iceland*

The breakwater at Bakkafjordur, a small fishing village in Northeast Iceland, is one of the first berm breakwaters worldwide. Constructed in 1983 and 1984 (Figure 9.28), it is 220 m long and extends to -7.0 m CD water depth at the outer part of the trunk and to 8.0 m CD at the breakwater roundhead, [Viggosson, 1990]. The 50-years design wave condition was $H_s = 4.8$ m with $T_p = 12.0$ s and design water level of +2.5 CD.

The design of the breakwater was based on the initial idea of a reshaping berm and was tested in a 3D physical model in a scale of 1:60. The berm on the trunk section was mostly constructed of 0.5–6.0 t armourstone with $M_{50} = 1.5$ t and stability number $H_{sD}/\Delta D_{n50} = 3.35$ (see Figure 9.29). According to the analysis presented in this book, this is beyond the acceptable level of $H_{sD}/\Delta D_{n50} = 3.0$ for a statically stable reshaping berm breakwater. As the seabed was hard, no apron was needed between the berm rock and the seabed.

The crest and inner berm were constructed of 2.0–6.0 t armourstone with an average mass of 3.0 t and a stability number $H_s/\Delta D_{n50} = 2.66$. On the breakwater head, a class of 6.0–10.0 t armourstone was used on top of the berm instead of the 2.0–6.0 t class, used on the trunk section.

This very early berm breakwater design is characterised by a steep berm front slope and compared to the geometrical design guidelines, both the berm level and crest level are low. The berm level is $0.3{*}H_{sD}$ and the crest level is $0.9{*}H_{sD}$ above design water level.

Figure 9.28. The Bakkafjordur breakwater and the local quarry in February 1986.

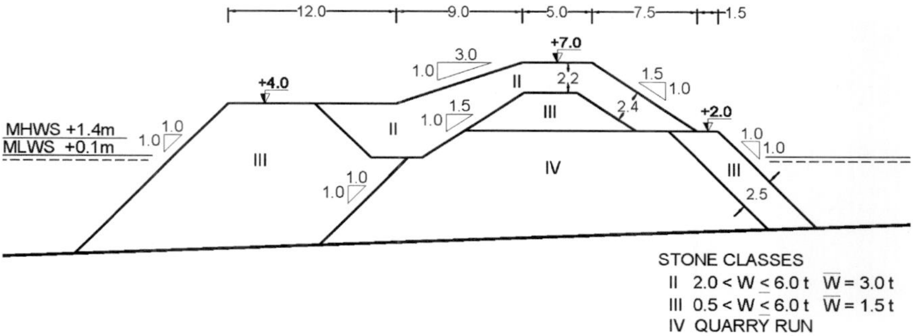

Figure 9.29. A design cross-section of the berm breakwater at Bakkafjordur from 1983, characterised by a steep front slope of the berm, low berm level and low crest level.

The breakwater consists of 105,000 m^3 of core and armourstones of rather poor quality quarried adjacent to the breakwater site, the Bakkafjordur quarry. During the construction period, problems regarding rock quality became evident. A new quarry was opened at Nonfell at a distance of 30 km from the site, where about 5,000 m^3 of armourstones larger than 2 t were obtained and used on the crest of the structure.

The rock in the Bakkafjordur quarry was highly altered basalt (Figure 9.30a) with a mass density of 2870 kg/m^3 and water absorption 0.9%. The rock from the Nonfell quarry was porphyritic basalt of Pleistocene age (Figure 9.30b) with a mass density of about 2850 kg/m^3 and water absorption between 1.0% and 1.5%. During the final inspection of the breakwater in December 1984, it became evident that some of the stones from the Nonfell quarry had split or broken in two or more pieces. Since then, further splitting or breakage of the in-place armourstone has not been observed.

There have been similar instances of splitting of armourstone shortly after quarrying during the winter time. If the rock from these quarries is allowed to acclimitise for a few weeks before being exposed to freezing condition, it appears to be less susceptible to splitting.

(a) (b)

Figure 9.30. (a) Highly altered basalt block from the Bakkafjordur quarry, and (b) slightly altered basalt block from the Nonfell quarry.

The Bakkafjordur breakwater was one of the first harbour projects in Iceland that was tendered out. Until then, most breakwaters had been constructed as in-house work by the Icelandic Harbour Authority. The tender was based on two designs, a conventional breakwater based on model tests from 1981, (Figure 9.31), and a berm breakwater design based on model tests from 1983. Eleven bids were received from ten contractors. The berm concept was chosen due to about 10% lower bids, although the total volume of stones was 23% higher.

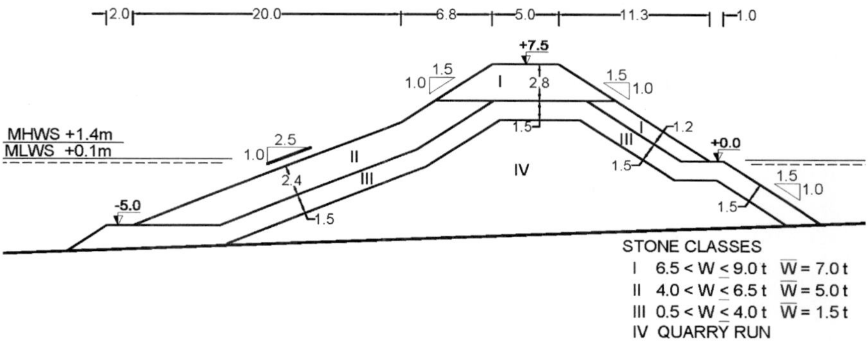

Figure 9.31. Alternative conventional cross-section for the Bakkafjordur breakwater.

Many problems came up during the construction of the breakwater. The quarry yield was not as good as what was indicated in the tender documents. The quality of the rock was poorer than what had been anticipated; and the contractor did not respect the lower size limits of the stone classes. According to the original contract the contractor should have completed the project in December 1983 but this was delayed by one year. The contractor remitted the case to arbitration where his claim was partly endorsed.

The construction technique in the first phase used dump trucks and bulldozers for placing the rocks [Viggosson, 1990]. This technique, however, resulted in too many fines that plugged voids and diminished the berm permeability. This method was therefore not accepted at the outer end of the breakwater where the berm was constructed using a backhoe excavator. A roadway was built over the top of the berm, which effectively filled all voids in the top surface of the berm. The roadway was only partly removed after construction. This caused contamination of the berm.

The reshaping of the main breakwater started shortly after construction; and in 1988, most of the berm had been eroded, as can be seen on the measured profile in Figure 9.32 and photo in Figure 9.33. The high stability number for the front part of the berm, $H_s/\Delta D_{n50} = 3.35$, as well as deterioration of stones in combination with contamination, has accelerated a dynamic development of the profile. Figure 9.30a, a photo taken in the summer of 1996, twelve years after construction, shows a stone from the Bakkafjordur quarry that was placed on the inner part of the breakwater head, where it had not moved. It is quite clear that this stone would split into many pieces should it start to move on the slope.

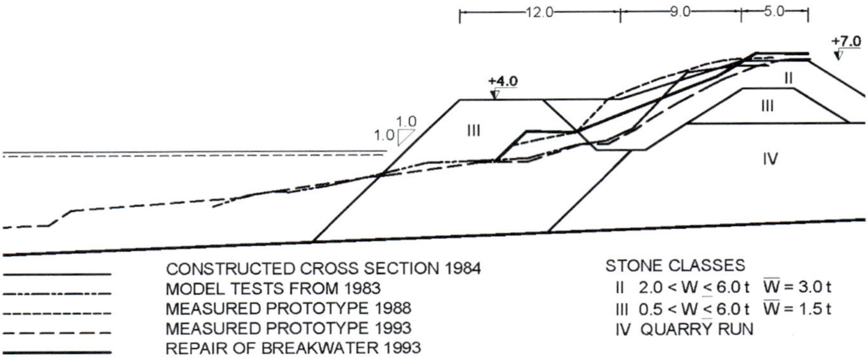

Figure 9.32. Cross-section of the Bakkafjordur breakwater, comparison between measured profiles from model tests in 1983 and monitoring of the breakwater in 1988 and 1993. The repair in 1993 has also been plotted. Note that compared to Figure 9.29, the stone classes here are numbered differently.

Figure 9.33. The Bakkafjordur breakwater in May 1988 with severe recession of the berm.

In the winter of 1992/93, the breakwater experienced a severe storm. The berm eroded up to the crest and an unstable S-profile had developed. Repair took place in 1993 and in spite of the poor quality of the rock, it was decided to use the local quarry again. The crest structure was rebuilt of stones larger than 2 t and the berm of stones from 0.5 to 2 t. Due to very limited funding, the repair of 1993 was marginal, seen in the dash-dot line on the cross-section in Figure 9.32. The original slope of the crest structure was not reached and the berm was much smaller, 2.5 m lower than the original berm, about half the width and of smaller stones. Figure 9.34 from the summer of 1995 shows that the small berm was already eroded at that time.

Figure 9.34. The Bakkafjordur harbour in the summer of 1995. A leeward breakwater was constructed in 1989.

In the autumn of 1995, the structure was again exposed by a storm, this time very close to the design wave conditions, [Sigurdarson *et al.*, 1998]. Photos and video recordings from the storm show that waves were breaking in front of and on the structure and heavy wave overtopping occurred (Figure 9.35). Local people regard the storm as one of the worst storms in that century, low barometric pressure, wind set-up and spring tide simultaneously.

Figure 9.35. Waves breaking in front of and over the Bakkafjordur breakwater during the storm of October 1995.

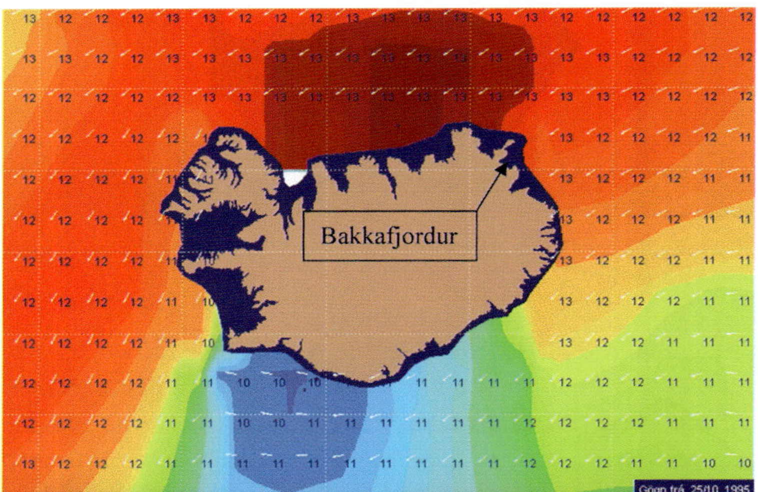

Figure 9.36. Wave hind cast (given in m) from the storm of 25 October 1995 at 12.00 hour from the European Centre for Medium-Range Weather Forecasts.

The offshore wave height estimated from a wave hindcast by ECMRWF gave H_s between 9 m and 10 m with T_m = 13 s (Figure 9.36). The breakwater was heavily damaged on the outermost 80 m, reaching to the inner side of the breakwater head. Again, the slope of the breakwater had eroded up to the crest. This time, only about half of the crest width remained, 1 m to 2 m or 1 to 2 rows of stones, in comparison with a 6 m width of the original design.

The structure was very close to breaching (Figure 9.37), but did not breach, which is remarkable. The material from the repair of 1993 and probably some of the original material had eroded and formed a 1:5 slope down to -4 m to -6 m CD water depth. Inspection of the reshaped profile showed that deterioration of the stones had caused filling and plugging of voids, reducing the porosity. The structure no longer functioned as a porous berm breakwater (Figure 9.38).

Figure 9.37. The Bakkafjordur breakwater reshaped after the storm of October 1995 into the crest, but did not breach.

The main conclusion that can be drawn from the Bakkafjordur breakwater is that in a dynamically stable structure, poor quality stones will break and the voids will gradually be filled up with smaller stones. This will decrease the ability of the structure to dissipate wave energy. Inspection at the site led to the conclusion that the poor quality (highly altered basalt) of the stones in the Bakkafjordur breakwater only accelerated a development that would occur over a longer time period, if it was built of better quality stones.

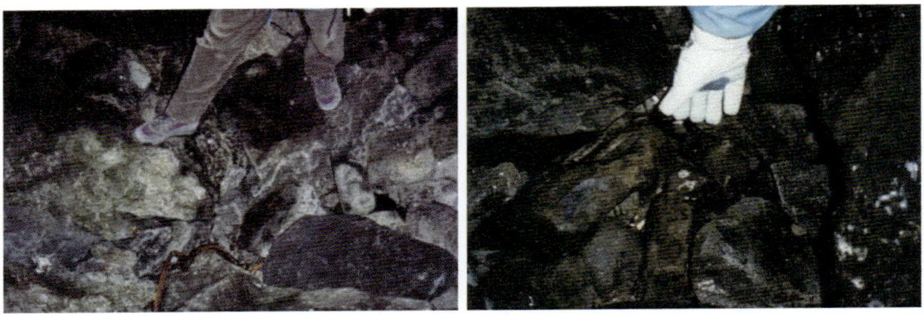

Figure 9.38. Inspection of the reshaped profile showed that deterioration of stones had caused filling and plugging of voids.

During the summer of 1996, the breakwater was repaired again. This time it was decided to take the stones from a quarry at Vopnafjordur, some 33 km south of Bakkafjordur. The rock was porphyritic basalt with a mass density of 2850 kg/m^3 and water absorption of 0.5%. The quality of the rock from Vopnafjordur is well known, as there is over 25 years of experience with a very exposed breakwater using that rock.

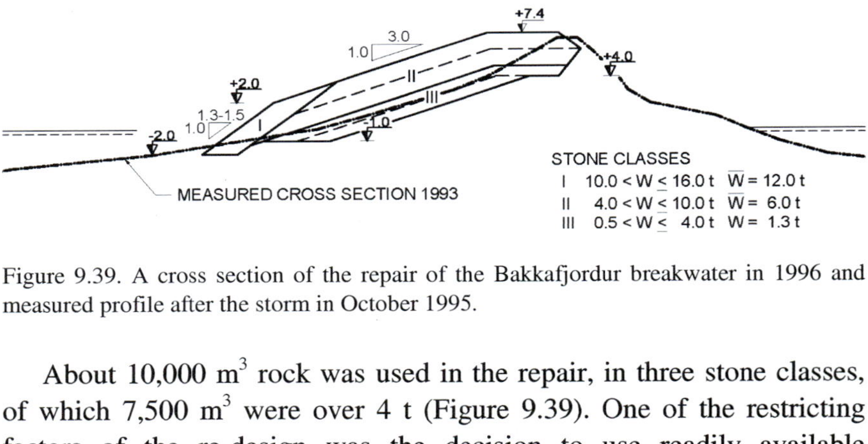

Figure 9.39. A cross section of the repair of the Bakkafjordur breakwater in 1996 and measured profile after the storm in October 1995.

About 10,000 m^3 rock was used in the repair, in three stone classes, of which 7,500 m^3 were over 4 t (Figure 9.39). One of the restricting factors of the re-design was the decision to use readily available equipment. This led to a design that made it possible to use 40–50 t excavators for the repair work. The design of the repair consisted of Class II armourstone, 4–10 t with a stability number $H_s/\Delta D_{n50} = 2.1$, in two layers with a slope of 1:3 from crest down to +2.0 m. In front of this, as a kind of toe protection, a single layer of Class I armourstone, 10–16 t with a stability number of 1.7, should be placed with a steep slope. The design of the repair fully utilised all material over 4 t. On the other hand, a large overproduction of smaller stones and quarry run was expected, of the order 15,000 m^3 to 20,000 m^3. This was anticipated in the contract as the contractor was to set aside this material in two separate mounds, quarry run and Class III stones 0.5 t to 4.0 t, to be used in planned harbour projects in Vopnafjordur. The repair work extended from the elbow and reached out to 120° on the breakwater head.

Through the years, the breakwater head has reshaped considerably. Small material and broken stones have been transported to the leeward side of the head to form a low mound, where the -1.0 m CD contour has moved about 10 m from its original position.

Inspection of the breakwater in October 1999, showed that the breakwater was mostly in good shape (Figure 9.40). Still some damage had developed on the elbow of the breakwater, where the repair from 1995 had started, and some of the Class I stones on the front of the repair had moved. In 2016, the breakwater was still in a good shape (Figure 9.41).

Figure 9.40. The Bakkafjordur breakwater in October 1999.

Figure 9.41. The Bakkafjordur breakwater in February 2016. Photo by Óli Þór Jakobsson.

9.5.3 *The Mortavika berm breakwater, Norway*

This section was written with the help of Arne E. Lothe, Norconsult, Norway.

The Mortavika berm breakwater in Norway protects a ferryboat harbour (Figure 9.42), and was built as a fully reshaping berm breakwater. It was constructed in 1991 to 1992 of tonalitic gneiss stones with a mass density of 2800 kg/m^3, quarried at the construction site, also visible at Figure 9.42. The design wave height at that time was estimated as $H_s = 6.8$ m with T_p = 15.6 s. The design was a multilayer berm structure with a steep lower slope of 1:1.3 and a gentle upper slope of 1:3 (Figure 9.43). The berm level is rather low, only 1 m above the design water level and as the tidal difference is very limited, the berm level is only 1.6 m above the mean low water spring tide. The initial width of the berm was 16 m, but as the upper slope is gentle, the armour width, A_h, was considerably larger at 30.6 m. The design process was presented in Lothe and Espedal, [1994].

The most exposed section is situated at a water depth of up to -21 m (Figure 9.44) which can be considered as a large water depth. The map shows that there is a steep and irregular foreshore. The berm was built of four armourstone classes, Class I in two layers on top of the berm with a median mass of 8 t. Below -1 m elevation the slope of the berm was protected by Classes II with a median mass of 5.5 t. Classes III and IV inside the berm have a median mass of 4 t and 2 t respectively

(Figure 9.45). With the estimated design wave height, this gives a stability number for Class I of 2.8, close to the design limit for fully reshaping berm breakwaters (Section 2.3).

Figure 9.42. General view of the Mortavika Breakwater and harbour after the 2005 storm. The rock was quarried from the paved area in the upper right of this image. Notice that the breakwater trunk is extensively reshaped, while the breakwater head, which is not as exposed, has retained its shape. Courtesy of Norwegian Public Roads Administration.

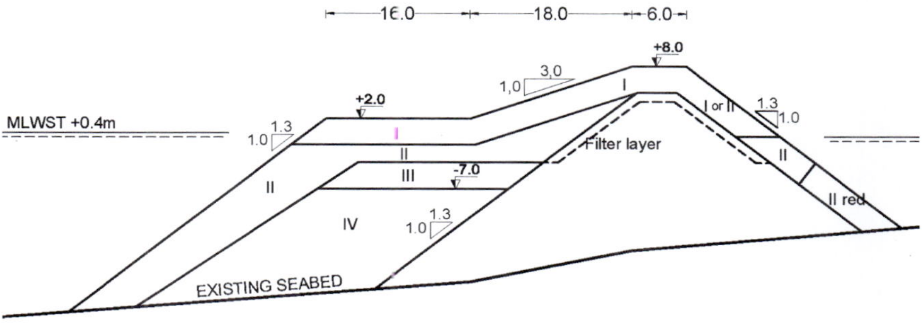

Figure 9.43. Cross section of the most exposed part of the Mortavika breakwater with a water depth of -21 m CD.

Figure 9.44. The Mortavika breakwater and bathymetry, clip from a nautical map. The most exposed part of the breakwater is the outer half of the trunk, which is also the deepest part. From web-map at geo.ngu.no/kart/arealis/.

Figure 9.45. Rocks on the berm of the Mortavika breakwater.

In the first winter, while under construction, the structure experienced a storm with an estimated significant wave height of 4.4 m, [Lothe *et al.* 2005]. About 100 m of the trunk section of the breakwater reshaped during the storm, with a recession of about 13 m, equal to about 9 armourstone diameters and 80% of the initial berm width. In 1993, about 3500 m^3 of rocks with M_{50}=8.0 t were "added to fill in anomalies detected during the completion survey". The originally intended berm width was only partially restored in the repair process, but the exact berm width after restoration is not known.

In January 2000, the breakwater experienced a storm with an estimated significant wave height of 4.1 m ($H_s/\Delta D_{n50}$ =1.7) and during inspection the breakwater was "found to be reshaped, but not damaged". In January 2005, the breakwater was hit by the heaviest storm in its life-time with estimated wave height of H_s=4.8 m ($H_s/\Delta D_{n50}$ =2.0). It was noted that "the berm was reshaped and at places entirely lost, and the crest was breached at one location in the middle of the breakwater". However, the term "breached" is not correct, as only little erosion occurred at the crest without lowering the crest itself, as seen in Figure 9.42 as well as Figure 9.46.

At this point it should be noted that the estimates of wave heights during these storms were based on transfer coefficients from the original wave refraction studies and the available offshore hindcast data that were used during the design of the breakwater.

A survey in March 2005 showed that the remaining primary berm width was reduced down to 0 m at places and typically to 1 m to 2 m. With an initial berm width of 16 m, the recession amounted to about 10 to 11D_{n50}, whereas for a stability number of $H_s/\Delta D_{n50}$ =2.0 a recession of only 1.6 D_{n50} would be expected, using Equation 3.19. This is a very large difference. The proposed repair strategy was to place 10,000 m^3 of rocks with M_{50} = 8.0 t on the reshaped profile.

Figure 9.46 View of the Mortavika breakwater after the January 2005 storms. The berm, originally at 16 m width, is lost closest to the camera, and rocks on the top at mid-length have been rearranged and removed. Courtesy of Norwegian Public Roads Administration.

Of course, the difference between the expected and real behaviour of the berm was noticed at that time and explanations were sought. Lothe and Lindefjeld, [2005] reported on the status of the breakwater after the storm in 2005. It was noted that an initial recession occurred in 1993, which may have been caused by the practice of "throwing" rocks into the berm using backhoe excavators. If the starting line for the recession is taken as the state of the repaired breakwater after the 1993 storm, then there is an excess recession, but not as dramatic as it would seem if the intended profile from the construction is taken as the starting point.

It was noted in Lothe and Lindefjeld [2005], however, that:

- There was poor agreement between the observed and the calculated recession;
- Large recession was observed for seemingly moderate wave loading.

Some possible explanations were offered, without giving a definitive answer. The hypotheses included such effects as; an adjustment of the breakwater centreline after the completion of the laboratory model studies; the breakwater being an early type with a wide and low berm and therefore possibly not conforming to modern formulae; deep-water wave effects not covered in the model; and insufficient stacking of the berm.

It may also be speculated that the rapid loss of the berm may be caused by rocks falling off an underwater cliff and not supporting the berm after their removal. In Section 3.7.3, it was proven that a deep toe berm, or no toe berm at all, will give a larger recession than with a toe berm at a certain depth (Figure 3.20). This argument may have some merit, but does not explain the total berm loss.

The most likely reason for the large recession was found recently when projects in the area provided an opportunity to re-examine the wave data. The original wave data estimates were made in 1989/1990, using the refraction models and source data (hindcast wave data) available at that time. These were used to determine the design wave height as well as the wave height estimates of experienced storms. Significant improvements in wave global and near-shore wave modelling have been made since then.

A recalculation of wave data based on offshore hindcast wave data from 1957–2014 shows that the 100-years return period significant wave height (in this case the design wave height) is probably H_{sD} = 7.9 m, which is a significant increase over the previous design wave height from 1991/92 of H_{sD} = 6.8 m. The recent data also show that the original design wave height has been equalled or exceeded on several occasions in the life-time of the breakwater. In the period 1990–2014 there have been 4 instances where the significant wave height has exceeded 6.5 m, the highest observation being H_s = 7.6 m (12 January 2005).

Using these new wave data, the design stability parameter of Class I rock becomes $H_s/\Delta D_{n50}$ = 3.4, and for the most severe storm in the breakwater life-time up to 2014 $H_s/\Delta D_{n50}$ = 3.2. This means that with these figures, the berm breakwater extends into the area of dynamic stability. Using Equation 3.19 gives a prediction of recession of 11.5D_{n50} or 16.5 m, and that is similar to the initial berm width of 16 m. The berm reshaping is therefore conforming to expectation, using the new wave data; although it was not intended to reshape as rapidly as has been the case in the period 1991–2014. This might be due to the exposed Class II rock of 5.5 t at the lower slope.

From this example of a berm breakwater it may be concluded that estimating the correct design conditions is extremely important.

9.5.4 *The Mackay Southern breakwater, Queensland, Australia*

The Mackay harbour in Queensland, Australia, was from the 1930s protected by the Southern and Northern breakwaters. In 1998 a mass-armoured berm breakwater was constructed to protect the Mackay Small Craft Harbour basin, adjacent to the Southern breakwater (Figure 9.47). It is a fully reshaping mass-armoured berm breakwater and has been maintained over time, following cyclones, by "topping-up" the breakwater profile with additional armour [Colleter *et al.*, 2011].

The breakwater was built with rock from the Mt Basset Quarry that has been used for construction and maintenance of breakwaters in Mackay since the 1930s. In the design phase of the Small Craft Harbour breakwater, a surface and subsurface quarry investigation was undertaken to determine the potential quarry yield in terms of blast block size distribution and to assess rock quality [Johnson *et al.*, 1999].

Figure 9.47. Mackay harbour, the Small Craft Harbour breakwater to the right on the photo protecting the marina that was built adjacent to the South breakwater. Photo by North Queensland Bulk Ports Corporation.

The design of the breakwater required an optimum yield of about 50% armour and 50% core and the adopted block size distribution separated armour and core at 30 kg, with armour heavier than 30 kg and core weighing less than 30 kg.

The breakwater was designed for various combinations of wave height, wave period and water level, with the most severe wave conditions being a 50-years return period wave event of H_s = 4.1 m, T_m = 9.0 s and still water level of +4.8 m LAT.

Figure 9.48 shows the grading curves for the estimated and actual quarry yield. The actual yield turned out to be considerably better than the predicted yield. Prior to its construction, it was estimated that less than 10% of the quarry yield would exceed 3 t, and this was also the largest rock size shown on the quarry yield curve. During the quarry development, the actual quarry yield above 3 t was over 30% with armourstone weighing up to 30 t. In addition, the median weight of the armour is close to 3 t, as estimated from the grading curve.

It should be noted that the actual quarry yield for the Mackay Small Craft Harbour breakwater between 3 t and 30 t is exactly the same as used for the construction of the Sirevåg breakwater in Norway.

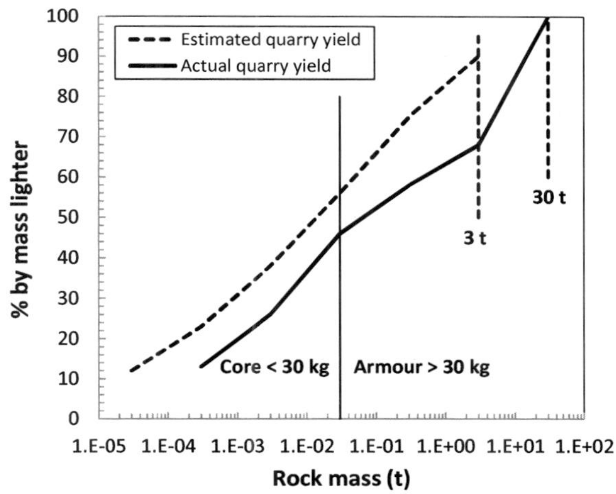

Figure 9.48. Grading curves for estimated and actual quarry yield for the Mackay Small Craft Harbour breakwater, re-drawn from Johnson *et al.* [1999].

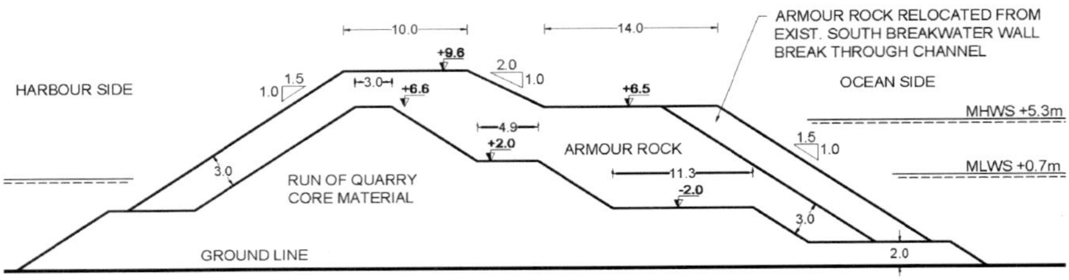

Figure 9.49. Design cross section of the Mackay Small Craft Harbour breakwater, from Johnson *et al.* [1999].

The adopted design cross-section is shown in Figure 9.49, [Johnson *et al.*, 1999]. The crest level was set at +9.6 m LAT to match with the existing Southern breakwater. The berm level was set at +6.5 m LAT, just above the highest astronomical tide. The idea was to build the berm at a level that makes the initial profile approximate the final S-shaped equilibrium profile as closely as possible. Based on 2D and 3D model tests, the adopted berm width was 14 m. While the tested cross-sections rapidly reshaped to form a long, flat, S-shaped profile, it was noted that larger rocks had been displaced to the toe of the profile whilst the smaller rocks remained on the flat part of the slope. The smaller rocks moved up the slope with each wave run-up and then down with each wave recession, gradually forming a dynamic equilibrium.

The breakwater was constructed with land-based construction equipment and most of the material was placed with end-tipping. This caused problems with segregation when larger rocks fell to the lower batter levels below LAT and smaller/finer material was deposited on the upper slopes in the tidal range and active wave zone. According to Johnson *et al.* [1999] this was rectified at least to some extent by a dragline, when a 6 m^3 bucket was used to pull larger rock from the lower batter up the slope and deposit the larger rock at the higher levels.

The breakwaters protecting the Mackay harbour, Northern, Southern and the Small Craft Harbour breakwaters, are mostly reshaping breakwaters and have needed year-to-year maintenance. Over 1997–2008, the total maintenance effort is estimated to be 60,000 t, corresponding to about 26 t per metre length on average.

During Tropical Cyclone Ului in March 2010, the breakwater was significantly damaged and substantial wave overtopping occurred, Colleter *et al.* (2011). The damage can be seen in Figure 9.50. At the peak of the storm the significant wave height along the breakwater varied between H_s = 4.0 m and 4.3 m. The reshaping extended well into the crest section with the bitumen sealed road on top of the breakwater damaged in areas. Rock up to 0.5 t and 1.0 t was left on the crest and the inner harbour revetment was damaged by wave overtopping. After the storm, the median rock mass on the emerged reshaped slope was estimated as 1.5 t with significant variations along its length. This corresponds to a stability number at the peak of the storm of $H_s/\Delta D_{n50}$ = 3.0 to 3.2.

Figure 9.50. The Mackay Small Craft Harbour breakwater after the Tropical Cyclone Ului [Colleter *et al.*, 2011]. Photo by Gildas Colleter.

Comparison between the Mackay Small Craft Harbour breakwater and the Sirevåg breakwater was presented in Sigurdarson *et al.*, [2012] (Table 9.5). Both breakwaters have been in service for a similar period, both were constructed from a dedicated armourstone quarry yielding about 30% in sizes 3 t to 30 t, and in both cases the aim of the design was to utilise all quarried material. Although the Sirevåg breakwater has been exposed to much larger waves than the Mackay Small Craft Harbour breakwater, the former did not need any repair, while the latter frequently needed maintenance—and recently, a major repair.

Table 9.5. Comparison between the Mackay Small Craft Harbour breakwater and the Sirevåg breakwater, wave load, construction material, experienced storms and maintenance need.

	Mackay Small Craft Harbour breakwater	*Sirevåg breakwater*
Type of berm breakwater	Mass-armoured, fully reshaping	Icelandic-type, partly reshaping
Construction year	1999	2001
Design wave height	$H_s = 4.1$ m	$H_s = 7.0$ m
Quarry yield 3 to 30 t	30%	30%
Main armour	One class, 30 kg-30 t	4 classes, 1–30 t
Cross sectional volume	Similar	Similar
Maintenance before major storm	Following cyclones "topping-up" with additional armour	None
Experienced storms close to design event	1 cyclone (2010): H_s= 4.0 to 4.3 m	storms (2002 and 2003): H_s= 7.0 and 7.1 m
Maintenance immediately after the storm	Initial maintenance of 60 t per m length	None

The reason for this huge difference lies in the design. In contrast to the Sirevåg breakwater, the design of the Mackay breakwater made no attempt to sort out the larger armourstone and place them in the most exposed areas of the cross section. As a result of segregation, both during construction as well as during the reshaping process, large armourstone ended up either at low levels or in the inner layers, where they are of little benefit. The extremely wide class of armour rock, that is all material between 30 kg and 30 t, makes the berm less permeable, and therefore reduces the breakwaters ability to absorb the incoming wave energy. With the design information in this book, a more economical and stable solution would have been designed.

Appendix A

Relationships to Compose a Damage Profile for a Straight Rock Slope

The parameters for a damage profile of a straight rock slope have been described in Section 3.5 and Figure 3.5. The relationships given below are valid for:

$H_oT_{om} < 120$ for $\cot\alpha \leq 3$ and $H_oT_{om} < 80 \cot\alpha - 120$ for $\cot\alpha > 3$ (A.1)

Parameters h_r, h_d, h_m and h_b, are given in Figure 3.5 and mark specific points on a straight slope, with respect to the still water level SWL.

The **parameter h_r** is given by:

$l_r/D_{n50}N_w^{0.2} = (0.065 - 0.06\ P) * (H_oT_{om} - 10 \cot\alpha)$ for $H_oT_{om} < 120$, (A.2)

and

$H_oT_{om} = (20 - 1.5 \cot\alpha) * l_r/D_{n50}N_w^{0.5} - 0.40$ for $H_oT_{om} \geq 120$ (A.3)

where $l_r = h_r \cot\alpha$

The **parameter h_d** is given by:

$h_d = h_r$ for: $S < 10$, or: $P = 0.1$ and $\cot\alpha \leq 3$, or: $P \neq 0.1$ and $\cot\alpha < 3$(A.4)

and $h_d/H_sN_w^{0.2} = 0.14 - (0.1\ P + 0.01) * (\cot\alpha - 3)$ (A.5)

for all other conditions.

The **parameter h_m** is given by:

$h_m/H_s = 0.15/s_{om}^{0.5}$ for $P \neq 0.6$ (A.6)

For P = 0.6 (a homogeneous structure):

for $\cot\alpha \leq 3$: $$h_m/H_s = 0.15/s_{om}^{0.5} \quad (A.7)$$

for $\cot\alpha \geq 5$: $$h_m/H_sN_w^{0.1} = 0.45\,\xi_m + 0.25 \text{ for } \xi_m < 4 \quad (A.8)$$

and: $$h_m/H_sN_w^{\ 0.1} = 2 \text{ for } \xi_m \geq 4 \quad (A.9)$$

For $3 < \cot\alpha < 5$ one should interpolate between $h_m/H_s = 0.15/s_{om}^{0.5}$ and h_b given below:

$$h_m(\text{interpolated}) = h_m + (h_b(\cot\alpha = 5) - h_m) * (\cot\alpha - 3)/2 \quad (A.10)$$

The **parameter h_b** is given by:

$$h_b/H_sN_w^{0.1} = 0.45\,\xi_m + 0.25 \qquad \text{for } \xi_m < 4 \quad (A.11)$$

$$h_b/H_sN_w^{0.1} = 2 \qquad \text{for } \xi_m \geq 4 \quad (A.12)$$

A spline can be constructed through the points on the slope with h_r, h_d, h_m and h_b, where the erosion area is predicted by the stability formulae 3.1 and 3.2. Accretion must be equal to erosion.

Appendix B

Detailed Analysis of Berm Recession

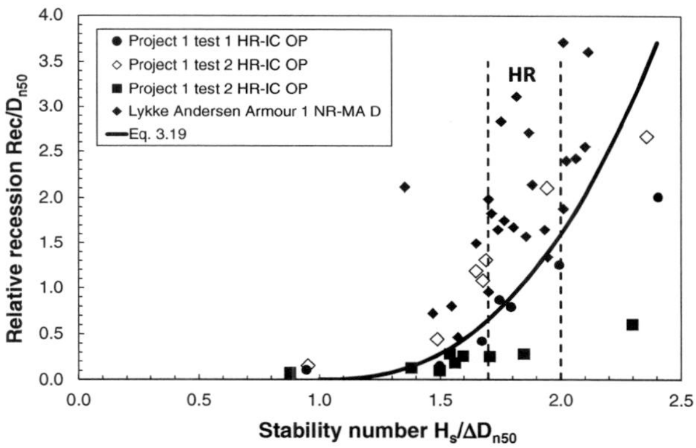

Figure B.1. Berm recession for all hardly reshaping berm breakwaters.

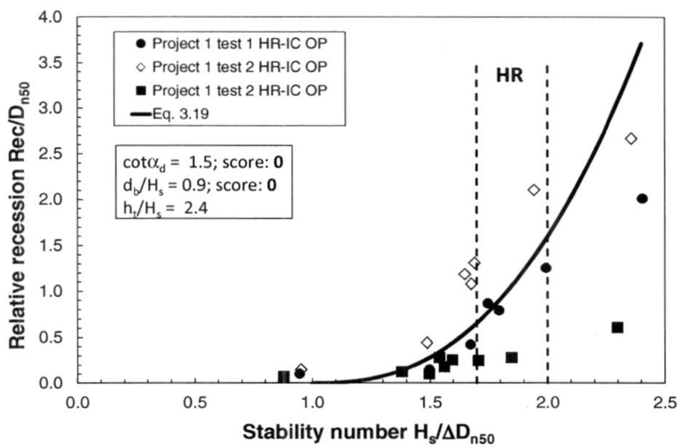

Figure B.2. Berm recession for Project 1.

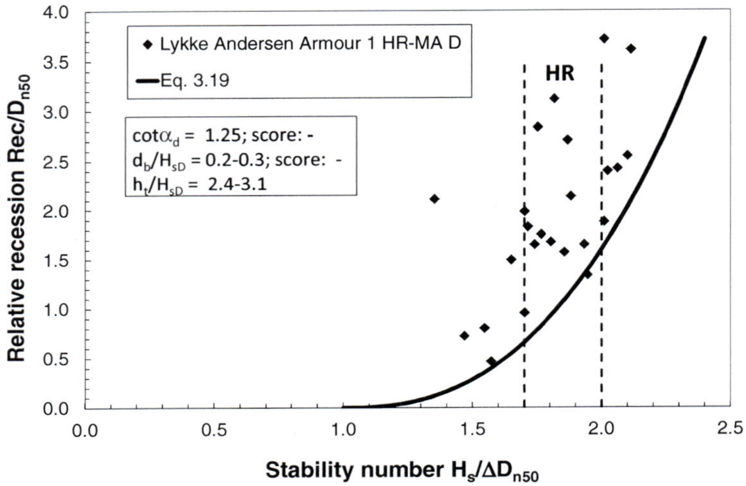

Figure B.3. Berm recession for Lykke Andersen (2006) Armour 1.

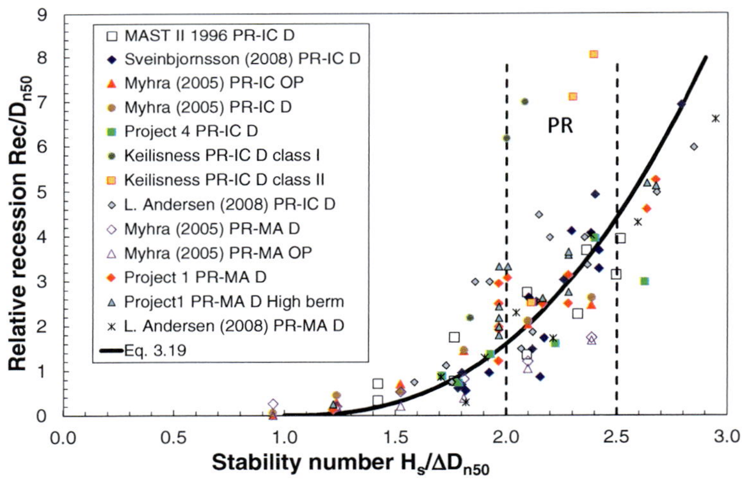

Figure B.4. Berm recession for all partly reshaping berm breakwaters.

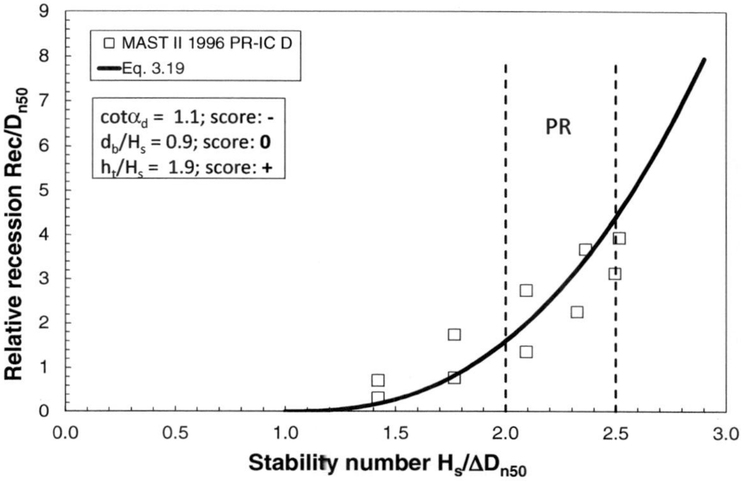

Figure B.5. Berm recession for MAST II.

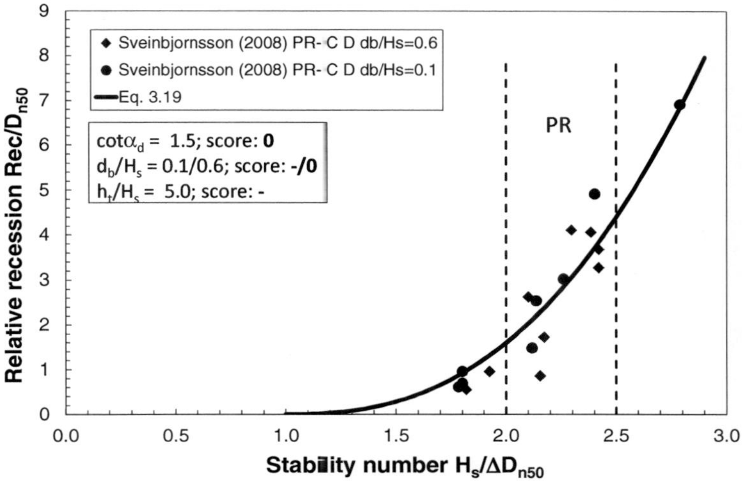

Figure B.6. Berm recession for Sveinbjörnsson (2008).

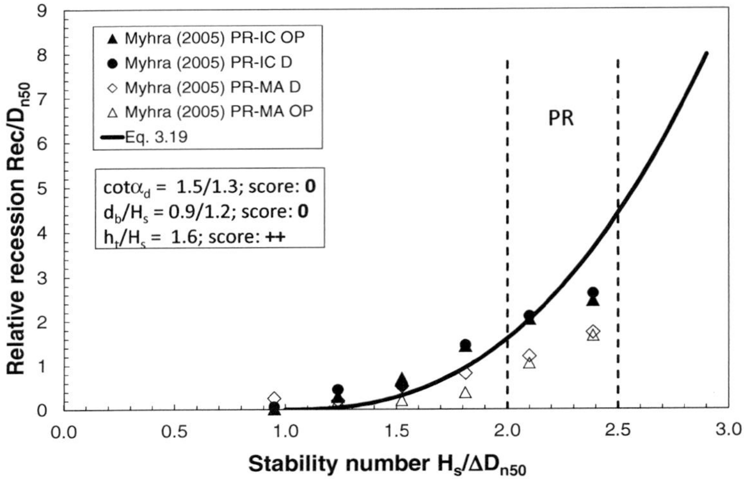

Figure B.7. Berm recession for Myhra (2005).

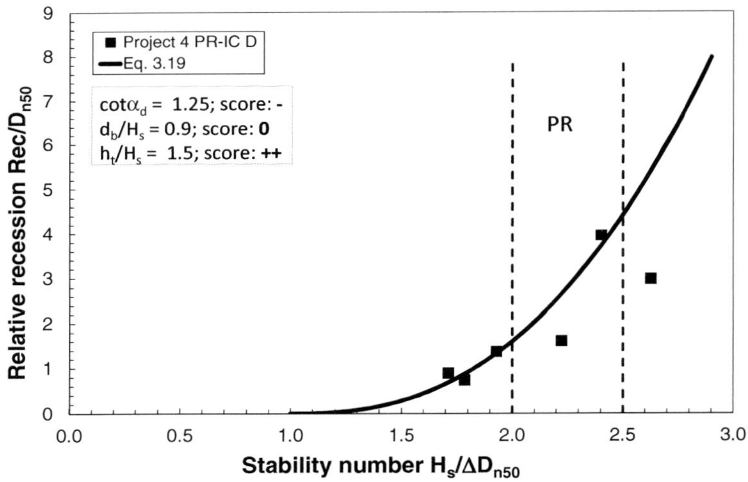

Figure B.8. Berm recession for Project 4.

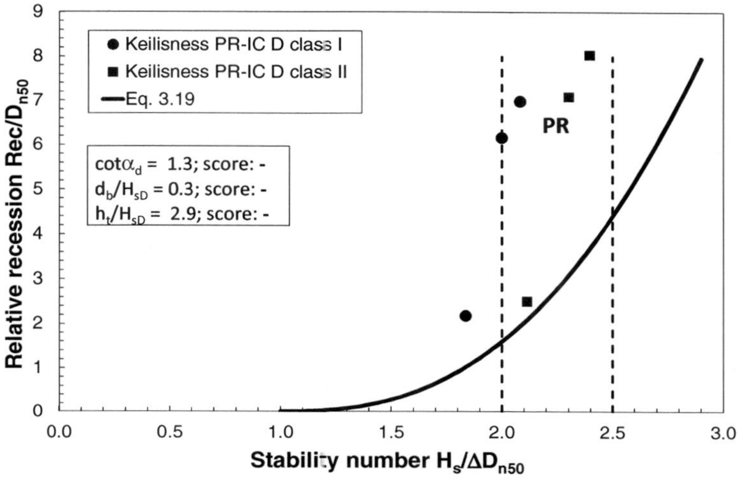

Figure B.9. Berm recession for Keilisness.

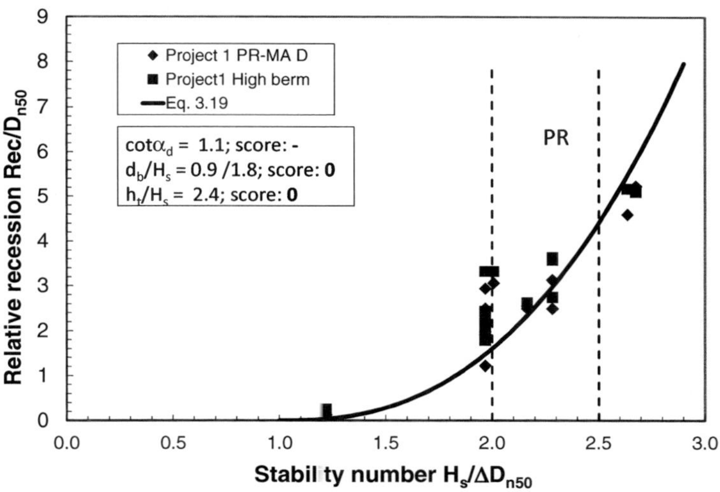

Figure B.10. Berm recession for Project 1.

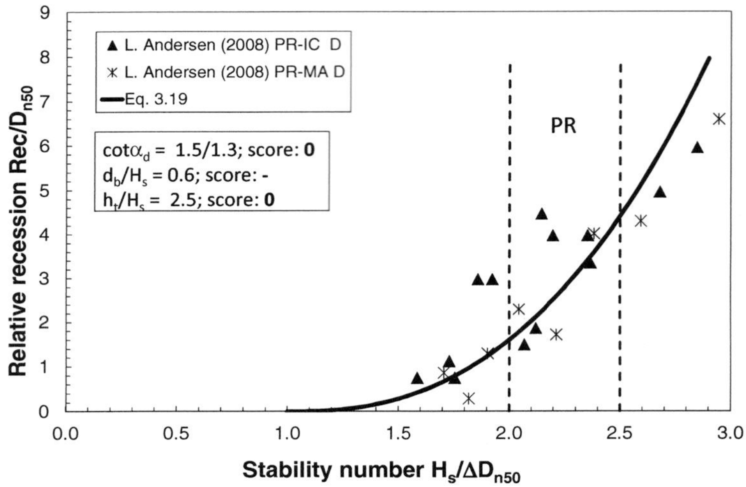

Figure B.11. Berm recession for Lykke Andersen *et al.* (2008).

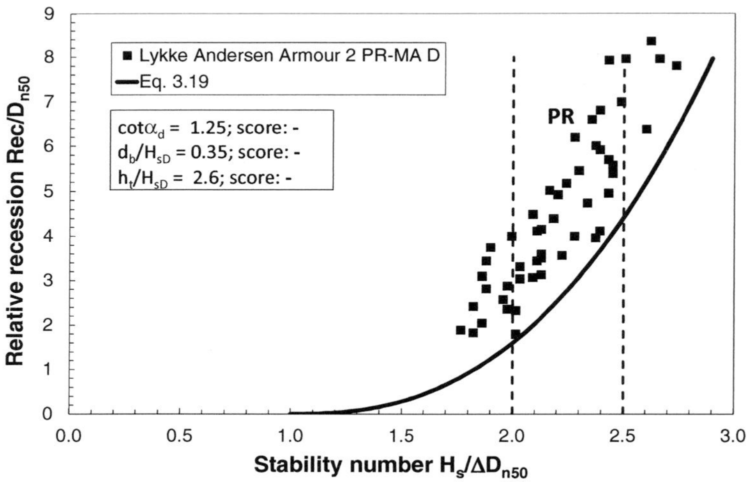

Figure B.12. Berm recession for Lykke Andersen (2006) Armour 2.

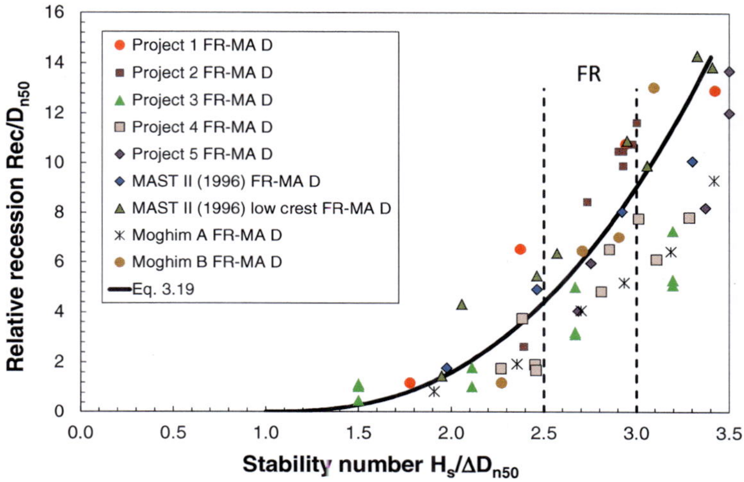

Figure B.13. Berm recession for all fully reshaping berm breakwaters.

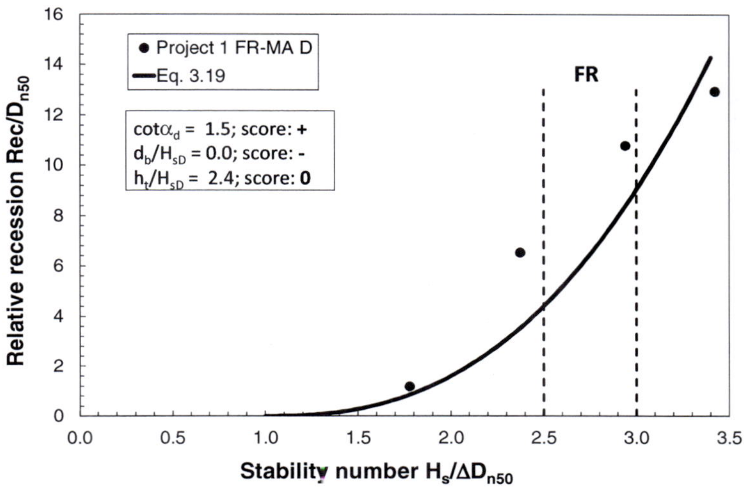

Figure B.14. Berm recession for Project 1.

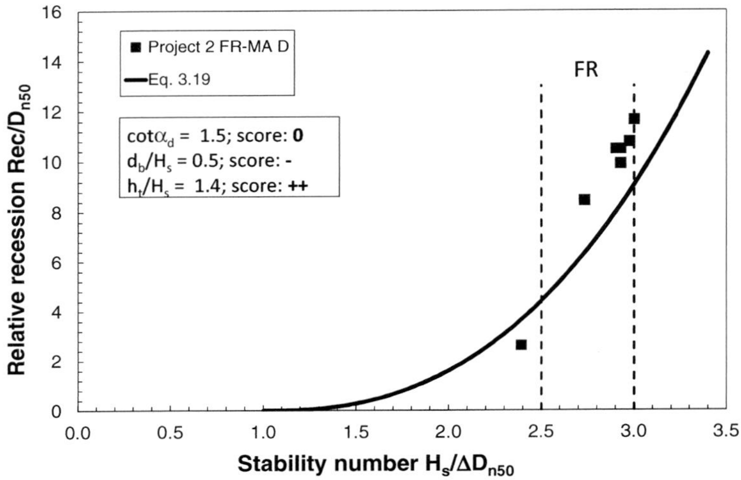

Figure B.15. Berm recession for Project 2.

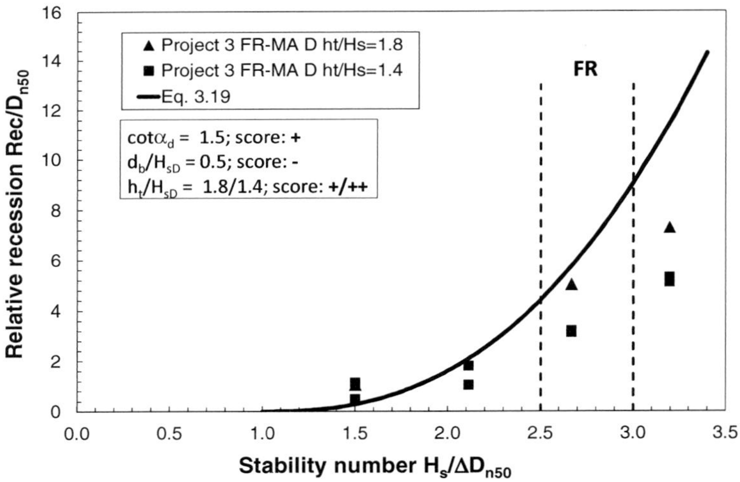

Figure B.16. Berm recession for Project 3.

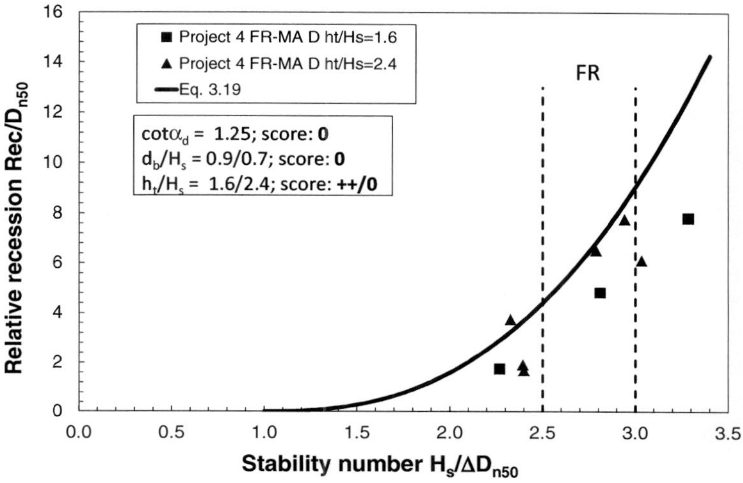

Figure B.17. Berm recession for Project 4.

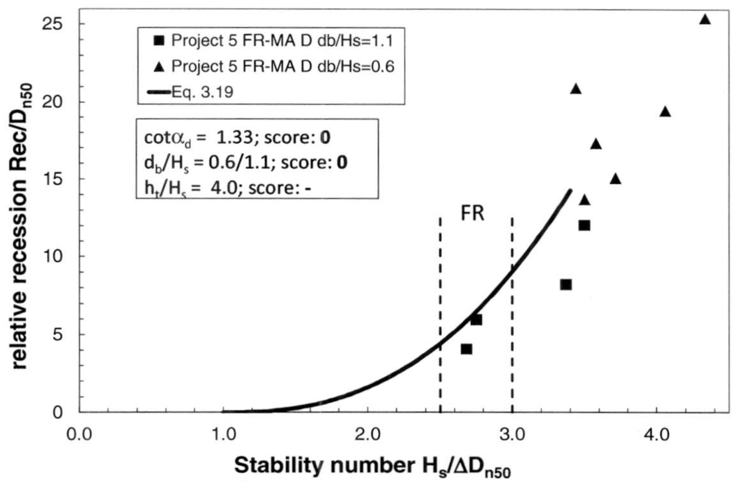

Figure B.18. Berm recession for Project 5.

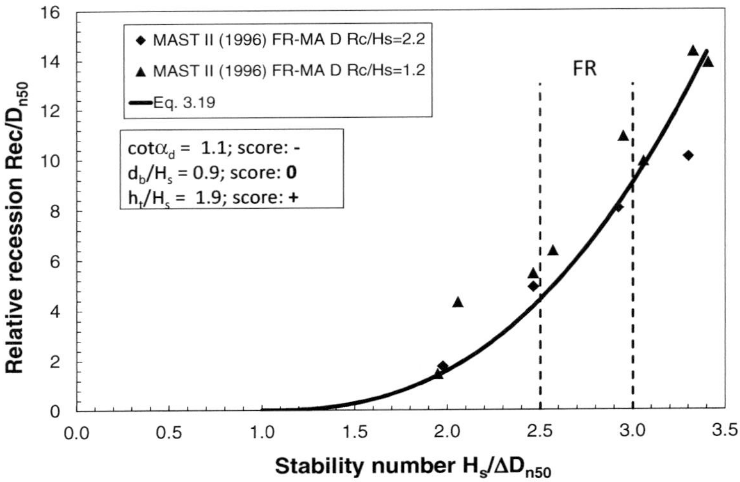

Figure B.19. Berm recession for MAST II.

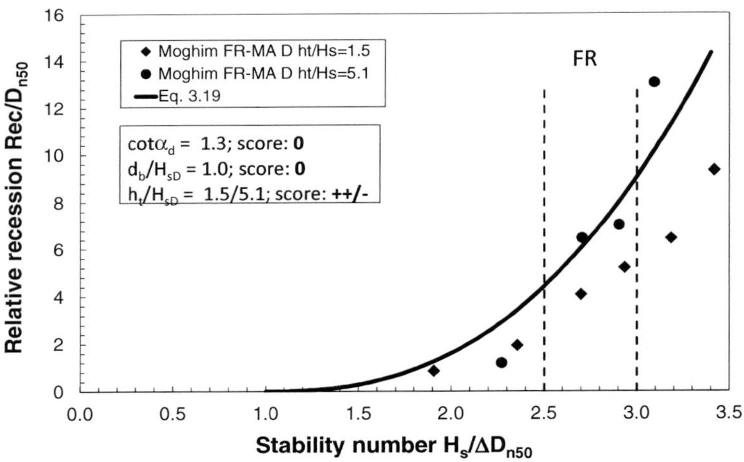

Figure B.20. Berm recession for Moghim (2009).

Appendix C

Detailed Analysis of Wave Overtopping

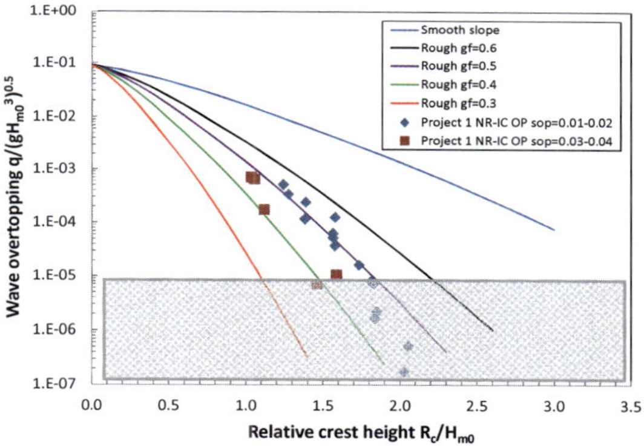

Figure C.1. Wave overtopping versus relative crest height. Project 1.

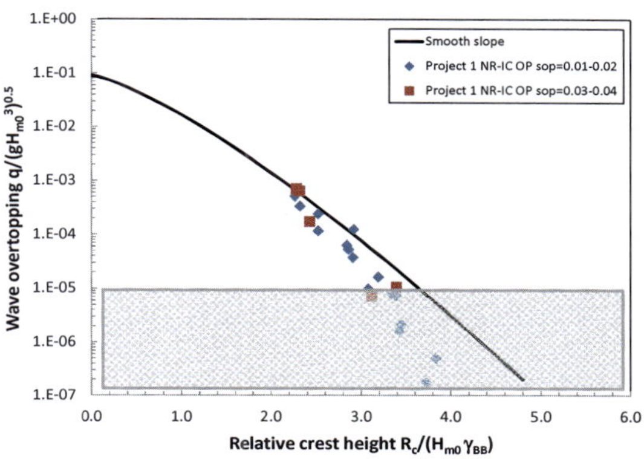

Figure C.2. Wave overtopping including γ_{BB}. Project 1.

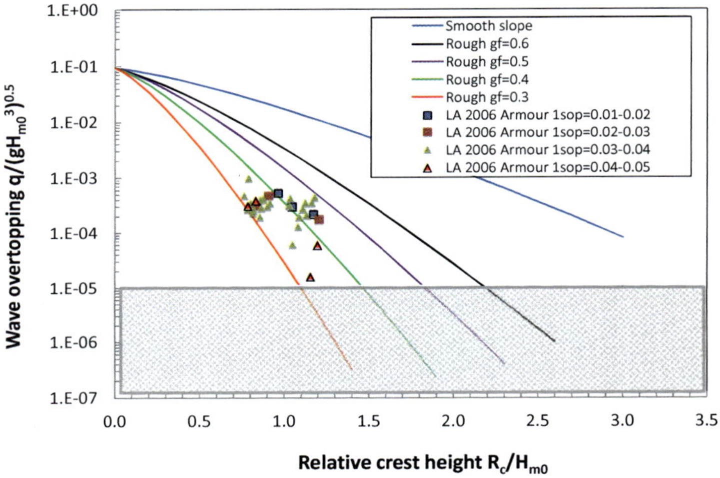

Figure C.3. Wave overtopping versus relative crest height. Lykke Andersen (2006) Armour 1.

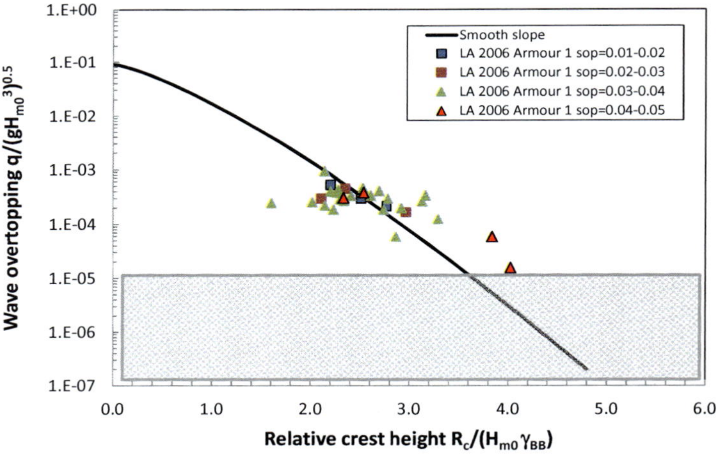

Figure C.4. Wave overtopping including γ_{BB}. Lykke Andersen (2006) Armour 1.

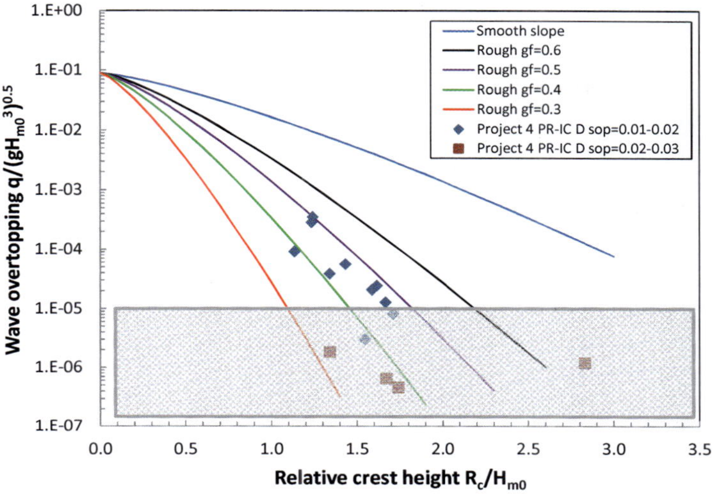

Figure C.5. Wave overtopping versus relative crest height. Project 4.

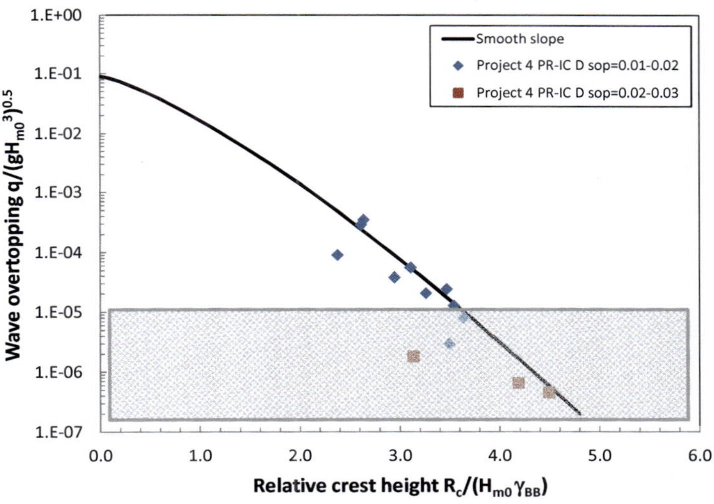

Figure C.6. Wave overtopping including γ_{BB}. Project 4.

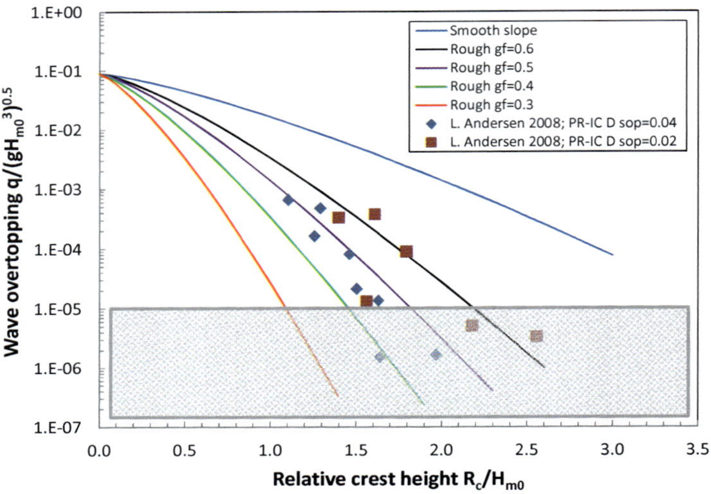

Figure C.7. Wave overtopping versus relative crest height. Lykke Andersen *et al.* (2008).

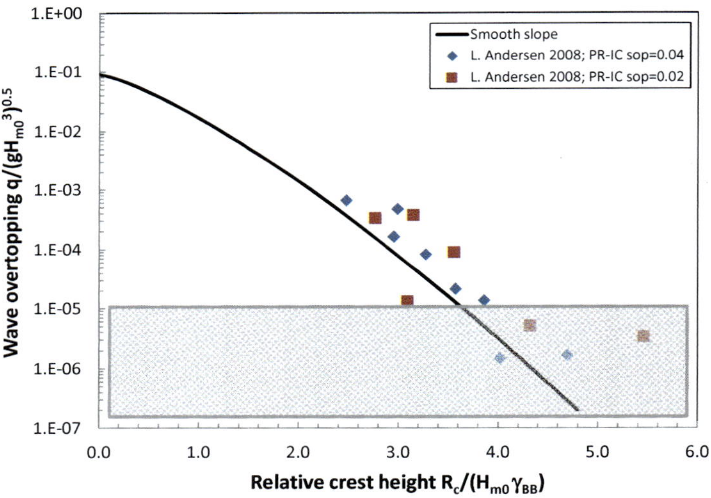

Figure C.8. Wave overtopping including γ_{BB}. Lykke Andersen *et al.* (2008).

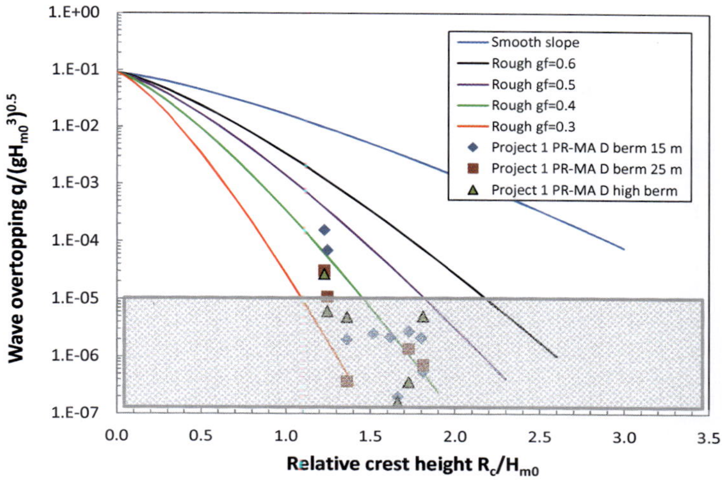

Figure C.9. Wave overtopping versus relative crest height. Project 1.

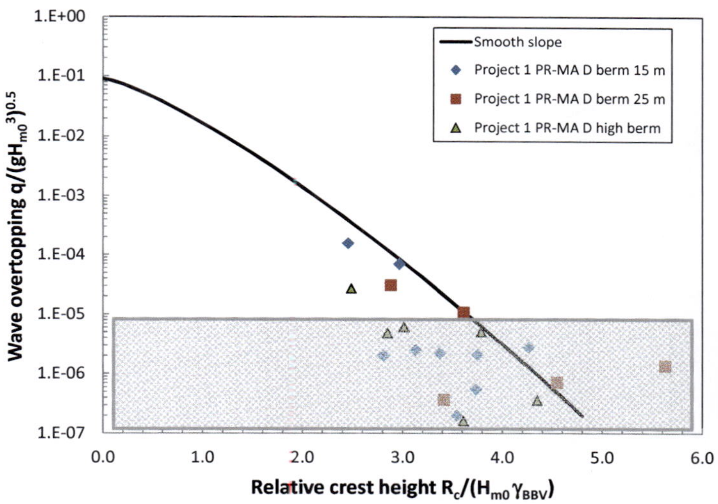

Figure C.10. Wave overtopping including γ_{BB}. Project 1.

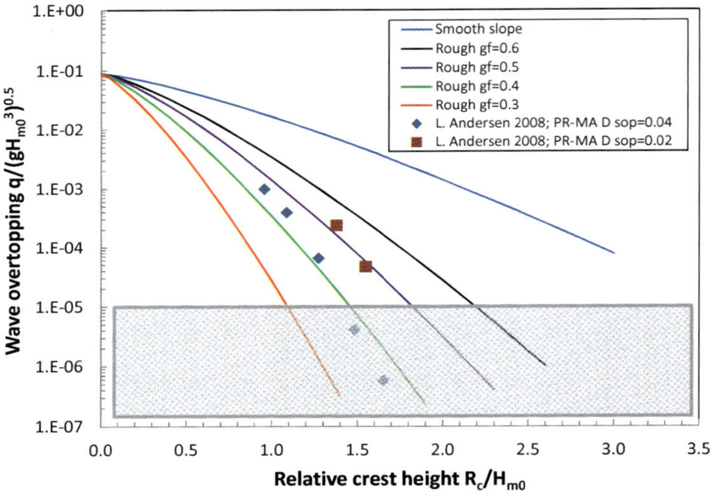

Figure C.11. Wave overtopping versus relative crest height. Lykke Andersen *et al.* (2008).

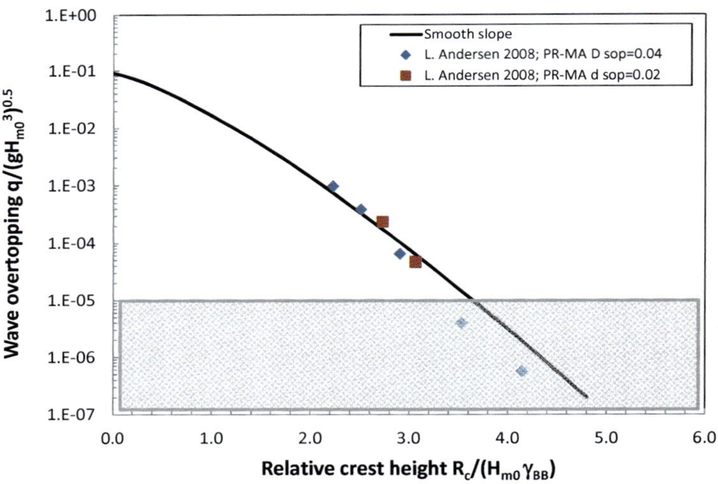

Figure C.12. Wave overtopping including γ_{BB}. Lykke Andersen *et al.* (2008).

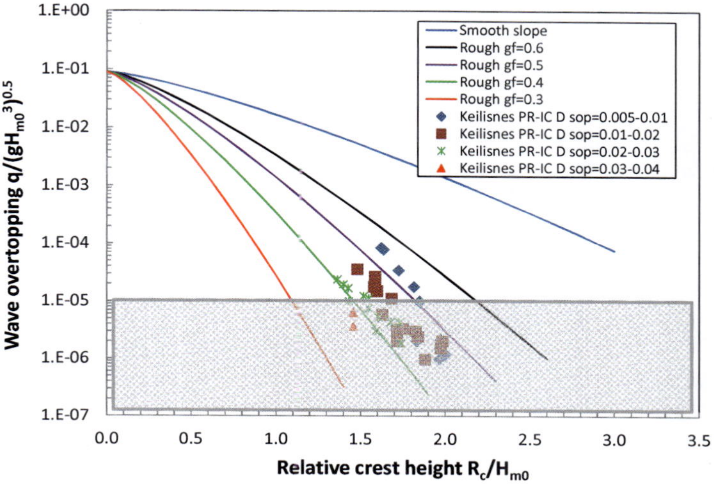

Figure C.13. Wave overtopping versus relative crest height. Keilisness.

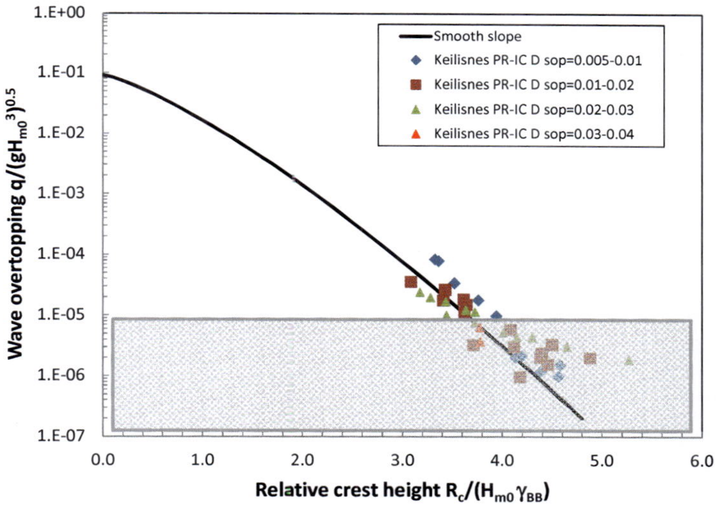

Figure C.14. Wave overtopping including γ_{BB}. Keilisness

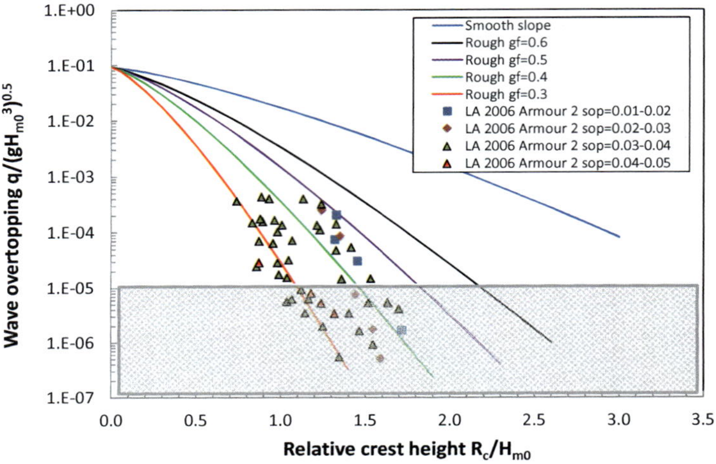

Figure C.15. Wave overtopping versus relative crest height. Lykke Andersen (2006) Armour 2.

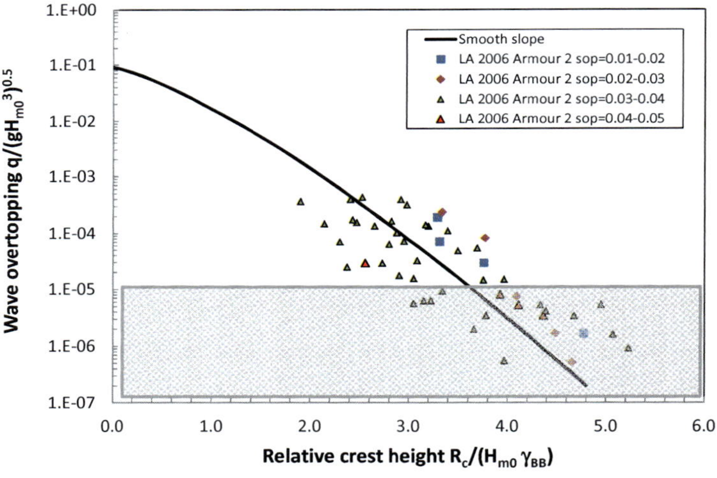

Figure C.16. Wave overtopping including γ_{BB}. Lykke Andersen (2006) Armour 2.

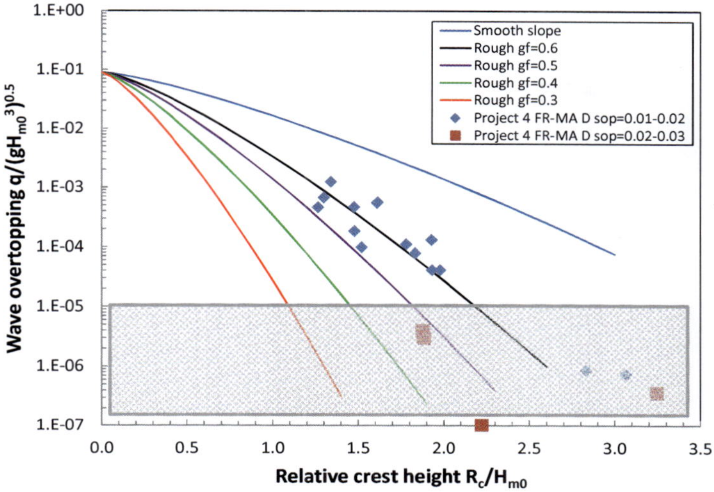

Figure C.17. Wave overtopping versus relative crest height. Project 4.

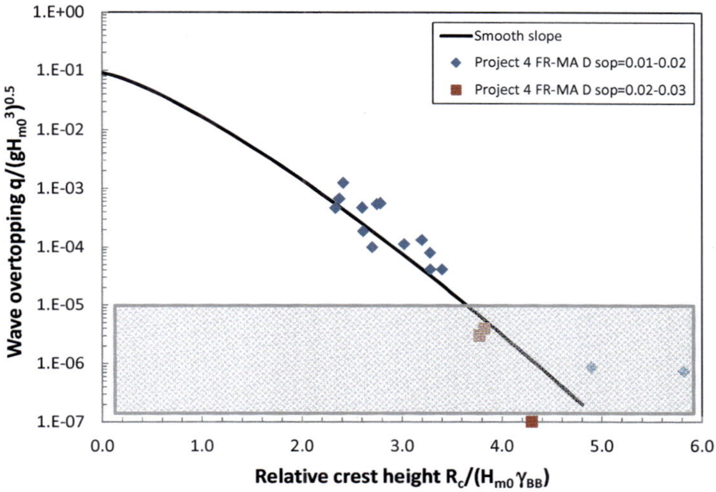

Figure C.18. Wave overtopping including γ_{BB}. Project 4.

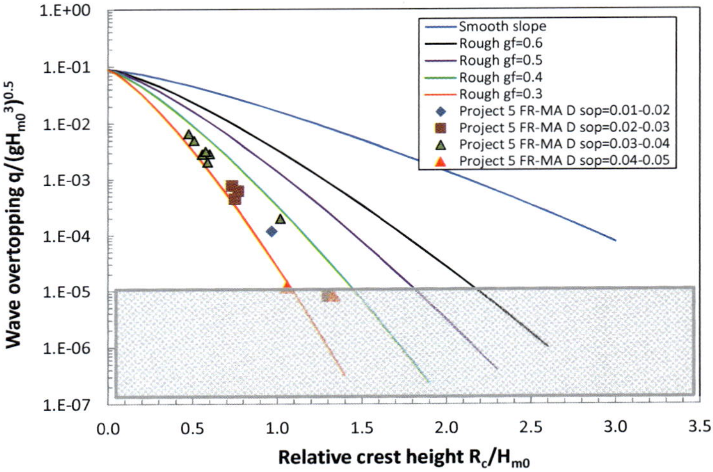

Figure C.19. Wave overtopping versus relative crest height. Project 5.

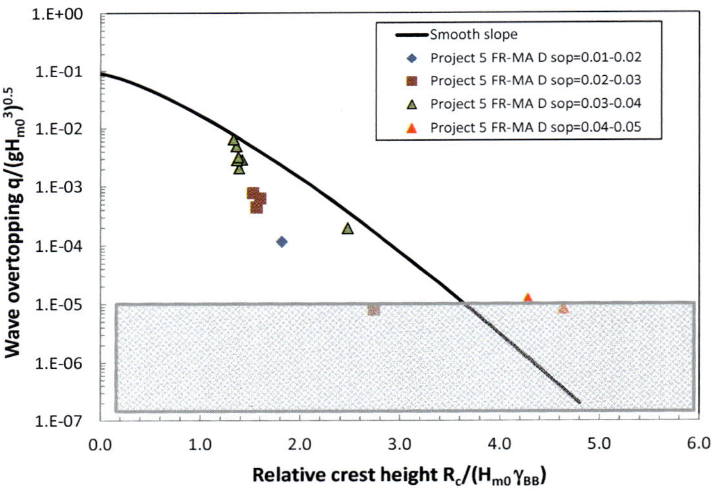

Figure C.20. Wave overtopping including γ_{BB}. Project 5.

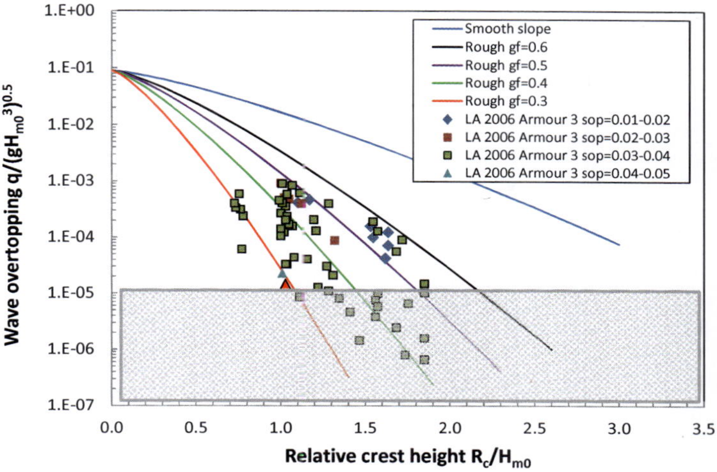

Figure C.21. Wave overtopping versus relative crest height. Lykke Andersen (2006) Armour 3.

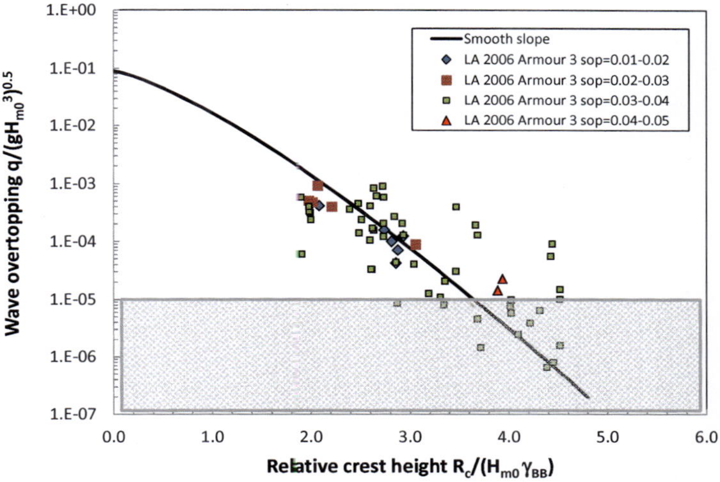

Figure C.22. Wave overtopping including γ_{BB}. Lykke Andersen (2006) Armour 3.

Appendix D

Calculations of Examples for Geometrical Design in Chapter 8

Section 8.2.1. HR IC dedicated quarry, Class I 10-20 t.

General conditions			**Outcome main parameters**			**Minimum transition level to class II**		
Design wave height H_{sD}	5	m	Wave steepness s_{op}	0.030	-	For H_{sD} at lowest level	-1.0	m CD
Peak period T_p	10.3	s	Relative mass density Δ	1.63	-	For lowest level with according H_s	-1.8	m CD
Overload H_s	6	m	Median mass Class I M_{50}	15.0	t	Design choice of transition for IC (3 rock classes)	-1.8	m CD
Design water level DWL	2	m CD	Nominal diameter Class I D_{n50}	1.77	m	Transition lower class for MA (2 rock classes)	-2.7	m CD
Lowest water level with H_{sD}	1	m CD	Stability number $H_{sD}/\Delta D_{n50}$	1.73	-			
Lowest storm level	0	m CD	**Type of berm breakwater**	**Hardly reshaping**		**Crest level ($\gamma_\beta = 1$)**		
H_s at lowest storm level	4.5	m	Number of rock classes for berm	3		If no overtopping criteria, $R_{c\,min}$	8.0	m CD
Mean High Water Spring	1	m CD	Basic recession for H_{sD} (no adaptation)	1.28	m	If no overtopping criteria, $R_{c\,max}$	9.0	m CD
Bottom level of foreshore at toe of structure	-10	m CD	Recession for overload (no adaptation)	2.82	m	For given allowable overtopping, q, γ_{BB}	0.46	
Allowable overtopping q for H_{sD}	1	l/s per m	Nominal diameter Class II, D_{n50}	1.37	m	Required crest level for design conditions	10.58	m CD
Allowable overtopping q for overload	10	l/s per m	Nominal diameter Class III, D_{n50}	0.97	m	Required crest level for overload	10.27	m CD
Mass density water	1025	kg/m^3				Design choice of crest level	10.00	m CD
Mass density rock	2700	kg/m^3	**Resiliency, berm width and level**					
			Wanted resiliency	15	%	**Check possibility of toe berm at level h_t**		
Choice of rock classes			Resulting Berm width B from resiliency	8.53	m	**Lowest** possible toe level (two layers)	-6.55	m CD
Rock Class I: minimum mass (0-15%)	10	t	Minimum berm width B_{min} from geometry	5.31	m	*Design conditions*		
Rock Class I: maximum mass (85-100%)	20	t	Berm level 0.6 H_{sD}	5	m CD	Allowable damage level for H_{sD}, N_{od}	2	-
Rock Class II: minimum mass (0-15%)	4	t	Δw for waves during construction	1	m	*Highest* level of toe for H_{sD} with chosen N_{od}	-4.22	m CD
Rock Class II: maximum mass (85-100%)	10	t	MHWS plus Δw = working level	2	m CD	Check validity range h_t/D_{n50}	5.4	ok
Rock Class III: M_{min} (leave open for MA)	1	t	Minimum berm level from construction	5.54	m CD	Check validity range h_t/h	0.47	ok
Rock Class III: M_{max} (leave open for MA)	4	t	Design choice of berm width	8.50	m	*Overload conditions*		
			Design choice of berm level	5.50	m CD	Allowable damage level for overload, N_{od}	4	-
						Highest level of toe for overload with chosen N_{od}	-4.72	m CD
			Required horizontal armour width A_h	17.3	m	Check validity range h_t/D_{n50}	5.9	ok
			Design choice of A_h	17.0	m	Check validity range h_t/h	0.52	ok
						Design choice of toe berm level (0 if no berm)	0	m CD
						Design choice cotα core below A_h	1.5	-

Section 8.2.2. PR IC standard gradings, Class I 6-10 t.

General conditions			**Outcome main parameters**			**Minimum transition level to class II**		
Design wave height H_{sD}	5	m	Wave steepness s_{op}	0.030	-	For H_{sD} at lowest level	-1.0	m CD
Peak period T_p	10.3	s	Relative mass density Δ	1.63	-	For lowest level with according H_s	-1.8	m CD
Overload H_s	6	m	Median mass Class I M_{50}	8.0	t	Design choice of transition for IC (3 rock classes)	-1.8	m CD
Design water level DWL	2	m CD	Nominal diameter Class I D_{n50}	1.44	m	Transition lower class for MA (2 rock classes)	-2.7	m CD
Lowest water level with H_{sD}	1	m CD	Stability number $H_{sD}/\Delta D_{n50}$	2.13	-			
Lowest storm level	0	m CD	**Type of berm breakwater**	**Partly reshaping**		**Crest level (γ_β = 1)**		
H_s at lowest storm level	4.5	m	Number of rock classes for berm	3		If no overtopping criteria, $R_{c\,min}$	8.0	m CD
Mean High Water Spring	1	m CD	Basic recession for H_{sD} (no adaptation)	3.12	m	If no overtopping criteria, $R_{c\,max}$	9.0	m CD
Bottom level of foreshore at toe of structure	-10	m CD	Recession for overload (no adaptation)	5.71	m	For given allowable overtopping, q, γ_{BB}	0.44	
Allowable overtopping q for H_{sD}	1	l/s per m	Nominal diameter Class II, D_{n50}	1.19	m	Required crest level for design conditions	10.20	m CD
Allowable overtopping q for overload	10	l/s per m	Nominal diameter Class III, D_{n50}	0.90	m	Required crest level for overload	9.91	m CD
Mass density water	1025	kg/m^3				Design choice of crest level	10.00	m CD
Mass density rock	2700	kg/m^3	**Resiliency, berm width and level**					
			Wanted resiliency	30	%	**Check possibility of toe berm at level h_t**		
Choice of rock classes			Resulting Berm width B from resiliency	10.40	m	**Lowest** possible toe level (two layers)	-6.69	m CD
Rock Class I: minimum mass (0-15%)	6	t	Minimum berm width B_{min} from geometry	4.56	m	*Design conditions*		
Rock Class I: maximum mass (85-100%)	10	t	Berm level 0.6 H_{sD}	5	m CD	Allowable damage level for H_{sD}, N_{od}	2	-
Rock Class II: minimum mass (0-15%)	3	t	Δw for waves during construction	1	m	*Highest* level of toe for H_{sD} with chosen N_{od}	-4.69	m CD
Rock Class II: maximum mass (85-100%)	6	t	MHWS plus Δw = working level	2	m CD	Check validity range h_t/D_{n50}	6.3	ok
Rock Class III: M_{min} (leave open for MA)	1	t	Minimum berm level from construction	4.87	m CD	Check validity range h_t/h	0.52	ok
Rock Class III: M_{max} (leave open for MA)	3	t	Design choice of berm width	10.50	m	*Overload conditions*		
			Design choice of berm level	5.00	m CD	Allowable damage level for overload, N_{od}	4	-
						Highest level of toe for overload with chosen N_{od}	-5.16	m CD
			Required horizontal armour width A_h	21.3	m	Check validity range h_t/D_{n50}	6.8	ok
			Design choice of A_h	21.0	m	Check validity range h_t/h	0.56	ok
						Design choice of toe berm level (0 if no berm)	-6	m CD
						Design choice cotα core below A_h	2	-

Section 8.2.3. FR MA standard gradings, Class I 3-6 t.

General conditions			**Outcome main parameters**			**Minimum transition level to class II**		
Design wave height H_{sD}	5	m	Wave steepness s_{op}	0.030	-	For H_{sD} at lowest level	-1.0	m CD
Peak period T_p	10.3	s	Relative mass density Δ	1.63	-	For lowest level with according H_s	-1.8	m CD
Overload H_s	6	m	Median mass Class I M_{50}	4.5	t	Design choice of transition for IC (3 rock classes)	-1.8	m CD
Design water level DWL	2	m CD	Nominal diameter Class I D_{n50}	1.19	m	Transition lower class for MA (2 rock classes)	-2.7	m CD
Lowest water level with H_{sD}	1	m CD	Stability number $H_{sD}/\Delta D_{n50}$	2.58	-			
Lowest storm level	0	m CD	**Type of berm breakwater**	**Fully reshaping**		**Crest level ($\gamma_\beta = 1$)**		
H_s at lowest storm level	4.5	m	Number of rock classes for berm	2		If no overtopping criteria, $R_{c\,min}$	8.0	m CD
Mean High Water Spring	1	m CD	Basic recession for H_{sD} (no adaptation)	5.96	m	If no overtopping criteria, $R_{c\,max}$	9.0	m CD
Bottom level of foreshore at toe of structure	-10	m CD	Recession for overload (no adaptation)	9.57	m	For given allowable overtopping, q, γ_{BB}	0.40	
Allowable overtopping q for H_{sD}	1	l/s per m	Nominal diameter Class II, D_{n50}	0.90	m	Required crest level for design conditions	9.42	m CD
Allowable overtopping q for overload	10	l/s per m	Nominal diameter Class III, D_{n50}	No Class		Required crest level for overload	9.15	m CD
Mass density water	1025	kg/m^3				Design choice of crest level	9.30	m CD
Mass density rock	2700	kg/m^3	**Resiliency, berm width and level**					
			Wanted resiliency	50	%	**Check possibility of toe berm at level h_t**		
Choice of rock classes			Resulting Berm width B from resiliency	11.92	m	**Lowest** possible toe level (two layers)	-6.69	m CD
Rock Class I: minimum mass (0-15%)	3	t	Minimum berm width B_{min} from geometry	7.14	m	*Design conditions*		
Rock Class I: maximum mass (85-100%)	6	t	Berm level 0.6 H_{sD}	5	m CD	Allowable damage level for H_{sD}, N_{od}	2	-
Rock Class II: minimum mass (0-15%)	1	t	Δw for waves during construction	1	m	*Highest* level of toe for H_{sD} with chosen N_{od}	-4.69	m CD
Rock Class II: maximum mass (85-100%)	3	t	MHWS plus Δw = working level	2	m CD	Check validity range h_t/D_{n50}	6.3	ok
Rock Class III: M_{min} (leave open for MA)		t	Minimum berm level from construction	4.37	m CD	Check validity range h_t/h	0.52	ok
Rock Class III: M_{max} (leave open for MA)		t	Design choice of berm width	12.00	m	*Overload conditions*		
			Design choice of berm level	4.50	m CD	Allowable damage level for overload, N_{od}	4	-
						Highest level of toe for overload with chosen N_{od}	-5.16	m CD
			Required horizontal armour width A_h	25.8	m	Check validity range h_t/D_{n50}	6.8	ok
			Design choice of A_h	26.0	m	Check validity range h_t/h	0.56	ok
						Design choice of toe berm level (0 if no berm)	0	m CD
						Design choice cotα core below A_h	2.5	-

Section 8.3.1. HR IC dedicated quarry, Class I 1–4 t.

General conditions			**Outcome main parameters**			**Minimum transition level to class II**		
Design wave height H_{sD}	3	m	Wave steepness s_{op}	0.020	-	For H_{sD} at lowest level	-0.2	m CD
Peak period T_p	9.8	s	Relative mass density Δ	1.54	-	For lowest level with according H_s	-1.2	m CD
Overload H_s	3.5	m	Median mass Class I M_{50}	2.5	t	Design choice of transition for IC (3 rock classes	-1.8	m CD
Design water level DWL	1	m CD	Nominal diameter Class I D_{n50}	0.99	m	Transition lower class for MA (2 rock classes)	-1.8	m CD
Lowest water level with H_{sD}	1	m CD	Stability number $H_{sD}/\Delta D_{n50}$	1.98	-			
Lowest storm level	0	m CD	**Type of berm breakwater**	**Hardly reshaping**		**Crest level ($\gamma_\beta = 1$)**		
H_s at lowest storm level	3	m	Number of rock classes for berm	2		If no overtopping criteria, $R_{c\,min}$	4.6	m CD
Mean High Water Spring	1	m CD	Basic recession for H_{sD} (no adaptation)	1.49	m	If no overtopping criteria, $R_{c\,max}$	5.2	m CD
Bottom level of foreshore at toe of structu	-9	m CD	Recession for overload (no adaptation)	2.28	m	For given allowable overtopping, q, γ_{BB}	0.46	
Allowable overtopping q for H_{sD}	5	l/s per m	Nominal diameter Class II, D_{n50}	0.61	m	Required crest level for design conditions	4.93	m CD
Allowable overtopping q for overload	20	l/s per m	Nominal diameter Class III, D_{n50}	No Class		Required crest level for overload	4.87	m CD
Mass density water	1025	kg/m^3				Design choice of crest level	4.80	m CD
Mass density rock	2600	kg/m^3	**Resiliency, berm width and level**					
			Wanted resiliency	20	%	**Check possibility of toe berm at level h_t**		
Choice of rock classes			Resulting Berm width B from resiliency	7.47	m	**Lowest** possible toe level (two layers)	-6.27	m CD
Rock Class I: minimum mass (0-15%)	1	t	Minimum berm width B_{min} from geomet	2.96	m	*Design conditions*		
Rock Class I: maximum mass (85-100%)	4	t	Berm level 0.6 H_{sD}	2.8	m CD	Allowable damage level for H_{sD}, N_{od}	2	-
Rock Class II: minimum mass (0-15%)	0.2	t	Δw for waves during construction	1	m	***Highest*** level of toe for H_{sD} with chosen N_{od}	-3.83	m CD
Rock Class II: maximum mass (85-100%)	1	t	MHWS plus Δw = working level	2	m CD	Check validity range h_t/D_{n50}	7.9	ok
Rock Class III: M_{min} (leave open for MA)		t	Minimum berm level from construction	3.97	m CD	Check validity range h_t/h	0.48	ok
Rock Class III: M_{max} (leave open for MA)		t	Design choice of berm width	8.00	m	*Overload conditions*		
			Design choice of berm level	4.00	m CD	Allowable damage level for overload, N_{od}	4	-
						Highest level of toe for overload with chosen N	-4.12	m CD
			Required horizontal armour width A_h	11.9	m	Check validity range h_t/D_{n50}	8.3	ok
			Design choice of A_h	12.0	m	Check validity range h_t/h	0.51	ok
						Design choice of toe berm level (0 if no berm)	0	m CD
						Design choice cotα core below A_h	1.5	-

Section 8.3.2. PR MA dedicated quarry, Class I 0.5–2 t.

General conditions			**Outcome main parameters**			**Minimum transition level to class II**		
Design wave height H_{sD}	3	m	Wave steepness s_{op}	0.020	-	For H_{sD} at lowest level	-0.2	m CD
Peak period T_p	9.8	s	Relative mass density Δ	1.54	-	For lowest level with according H_s	-1.2	m CD
Overload H_s	3.5	m	Median mass Class I M_{50}	1.3	t	Design choice of transition for IC (3 rock classes)	-1.8	m CD
Design water level DWL	1	m CD	Nominal diameter Class I D_{n50}	0.78	m	Transition lower class for MA (2 rock classes)	-1.8	m CD
Lowest water level with H_{sD}	1	m CD	Stability number $H_{sD}/\Delta D_{n50}$	2.49	-			
Lowest storm level	0	m CD	**Type of berm breakwater**	**Partly reshaping**		**Crest level (γ_β = 1)**		
H_s at lowest storm level	3	m	Number of rock classes for berm	2		If no overtopping criteria, $R_{c\,min}$	4.6	m CD
Mean High Water Spring	1	m CD	Basic recession for H_{sD} (no adaptation)	3.41	m	If no overtopping criteria, $R_{c\,max}$	5.2	m CD
Bottom level of foreshore at toe of structure	-9	m CD	Recession for overload (no adaptation)	4.49	m	For given allowable overtopping, q, γ_{BB}	0.34	
Allowable overtopping q for H_{sD}	5	l/s per m	Nominal diameter Class II, D_{n50}	0.49	m	Required crest level for design conditions	3.92	m CD
Allowable overtopping q for overload	20	l/s per m	Nominal diameter Class III, D_{n50}	No Class		Required crest level for overload	3.87	m CD
Mass density water	1025	kg/m^3				Design choice of crest level	4.50	m CD
Mass density rock	2600	kg/m^3	**Resiliency, berm width and level**					
			Wanted resiliency	30	%	**Check possibility of toe berm at level h_t**		
Choice of rock classes			Resulting Berm width B from resiliency	11.36	m	**Lowest** possible toe level (two layers)	-6.53	m CD
Rock Class I: minimum mass (0-15%)	0.5	t	Minimum berm width B_{min} from geometry	4.19	m	*Design conditions*		
Rock Class I: maximum mass (85-100%)	2	t	Berm level 0.6 H_{sD}	2.8	m CD	Allowable damage level for H_{sD}, N_{od}	2	-
Rock Class II: minimum mass (0-15%)	0.1	t	Δw for waves during construction	1	m	*Highest* level of toe for H_{sD} with chosen N_{od}	-5.08	m CD
Rock Class II: maximum mass (85-100%)	0.5	t	MHWS plus Δw = working level	2	m CD	Check validity range h_t/D_{n50}	12.5	ok
Rock Class III: M_{min} (leave open for MA)		t	Minimum berm level from construction	3.57	m CD	Check validity range h_t/h	0.61	ok
Rock Class III: M_{max} (leave open for MA)		t	Design choice of berm width	11.50	m	*Overload conditions*		
			Design choice of berm level	2.80	m CD	Allowable damage level for overload, N_{od}	4	-
						Highest level of toe for overload with chosen N_{od}	-5.33	m CD
			Required horizontal armour width A_h	15.0	m	Check validity range h_t/D_{n50}	13.0	ok
			Design choice of A_h	15.0	m	Check validity range h_t/h	0.63	ok
						Design choice of toe berm level (0 if no berm)	-5.5	m CD
						Design choice cotα core below A_h	1.5	-

Section 8.4.1. HR IC dedicated quarry, Class I 20–35 t.

General conditions			**Outcome main parameters**			**Minimum transition level to class II**		
Design wave height H_{sD}	7	m	Wave steepness s_{op}	0.040	-	For H_{sD} at lowest level	-1.8	m CD
Peak period T_p	10.6	s	Relative mass density Δ	1.62	-	For lowest level with according H_s	-2.6	m CD
Overload H_s	8	m	Median mass Class I M_{50}	27.5	t	Design choice of transition for IC (3 rock classes)	-1.8	m CD
Design water level DWL	4	m CD	Nominal diameter Class I D_{n50}	2.17	m	Transition lower class for MA (2 rock classes)	-3.9	m CD
Lowest water level with H_{sD}	1	m CD	Stability number $H_{sD}/\Delta D_{n50}$	1.99	-			
Lowest storm level	0	m CD	**Type of berm breakwater**	**Hardly reshaping**		**Crest level (γ_β = 1)**		
H_s at lowest storm level	6.5	m	Number of rock classes for berm	3		If no overtopping criteria, $R_{c\,min}$	12.4	m CD
Mean High Water Spring	2	m CD	Basic recession for H_{sD} (no adaptation)	3.40	m	If no overtopping criteria, $R_{c\,max}$	13.8	m CD
Bottom level of foreshore at toe of structure	-18	m CD	Recession for overload (no adaptation)	6.87	m	For given allowable overtopping, q, γ_{BB}	0.38	
Allowable overtopping q for H_{sD}	10	l/s per m	Nominal diameter Class II, D_{n50}	1.77	m	Required crest level for design conditions	12.19	m CD
Allowable overtopping q for overload	no	l/s per m	Nominal diameter Class III, D_{n50}	1.37	m	Required crest level for overload		m CD
Mass density water	1030	kg/m^3				Design choice of crest level	12.50	m CD
Mass density rock	2700	kg/m^3	**Resiliency, berm width and level**					
			Wanted resiliency	20	%	**Check possibility of toe berm at level h_t**		
Choice of rock classes			Resulting Berm width B from resiliency	16.98	m	**Lowest** possible toe level (two layers)	-13.75	m CD
Rock Class I: minimum mass (0-15%)	20	t	Minimum berm width B_{min} from geometry	6.50	m	*Design conditions*		
Rock Class I: maximum mass (85-100%)	35	t	Berm level 0.6 H_{sD}	8.2	m CD	Allowable damage level for H_{sD}, N_{od}	2	-
Rock Class II: minimum mass (0-15%)	10	t	Δw for waves during construction	1	m	*Highest* level of toe for H_{sD} with chosen N_{od}	-8.03	m CD
Rock Class II: maximum mass (85-100%)	20	t	MHWS plus Δw = working level	3	m CD	Check validity range h_t/D_{n50}	6.6	ok
Rock Class III: M_{min} (leave open for MA)	4	t	Minimum berm level from construction	7.34	m CD	Check validity range h_t/h	0.48	ok
Rock Class III: M_{max} (leave open for MA)	10	t	Design choice of berm width	17.00	m	*Overload conditions*		
			Design choice of berm level	7.50	m CD	Allowable damage level for overload, N_{od}	4	-
						Highest level of toe for overload with chosen N_{od}	-8.36	m CD
			Required horizontal armour width A_h	27.9	m	Check validity range h_t/D_{n50}	6.8	ok
			Design choice of A_h	28.0	m	Check validity range h_t/h	0.49	ok
						Design choice of toe berm level (0 if no berm)	-8	m CD
						Design choice cotα core below A_h	1.5	-

Section 8.4.2. PR IC dedicated quarry, Class I 10-20 t.

General conditions			**Outcome main parameters**			**Minimum transition level to class II**		
Design wave height H_{sD}	7	m	Wave steepness s_{op}	0.040	-	For H_{sD} at lowest level	-1.8	m CD
Peak period T_p	10.6	s	Relative mass density Δ	1.62	-	For lowest level with according H_s	-2.6	m CD
Overload H_s	8	m	Median mass Class I M_{50}	15.0	t	Design choice of transition for IC (3 rock classes)	-2.6	m CD
Design water level DWL	4	m CD	Nominal diameter Class I D_{n50}	1.77	m	Transition lower class for MA (2 rock classes)	-3.9	m CD
Lowest water level with H_{sD}	1	m CD	Stability number $H_{sD}/\Delta D_{n50}$	2.44	-			
Lowest storm level	0	m CD	**Type of berm breakwater**	**Partly reshaping**		**Crest level (γ_β = 1)**		
H_s at lowest storm level	6.5	m	Number of rock classes for berm	3		If no overtopping criteria, $R_{c\,min}$	12.4	m CD
Mean High Water Spring	2	m CD	Basic recession for H_{sD} (no adaptation)	7.02	m	If no overtopping criteria, $R_{c\,max}$	13.8	m CD
Bottom level of foreshore at toe of structure	-18	m CD	Recession for overload (no adaptation)	12.26	m	For given allowable overtopping, q, γ_{BB}	-0.22	
Allowable overtopping q for H_{sD}	10	l/s per m	Nominal diameter Class II, D_{n50}	1.37	m	Required crest level for design conditions	-0.80	m CD
Allowable overtopping q for overload	no	l/s per m	Nominal diameter Class III, D_{n50}	0.97	m	Required crest level for overload		m CD
Mass density water	1030	kg/m^3				Design choice of crest level	12.50	m CD
Mass density rock	2700	kg/m^3	**Resiliency, berm width and level**					
			Wanted resiliency	30	%	**Check possibility of toe berm at level h_t**		
Choice of rock classes			Resulting Berm width B from resiliency	23.41	m	**Lowest** possible toe level (two layers)	-14.55	m CD
Rock Class I: minimum mass (0-15%)	10	t	Minimum berm width B_{min} from geometry	8.79	m	*Design conditions*		
Rock Class I: maximum mass (85-100%)	20	t	Berm level 0.6 H_{sD}	8.2	m CD	Allowable damage level for H_{sD}, N_{od}	2	-
Rock Class II: minimum mass (0-15%)	4	t	Δw for waves during construction	1	m	*Highest* level of toe for H_{sD} with chosen N_{od}	-11.48	m CD
Rock Class II: maximum mass (85-100%)	10	t	MHWS plus Δw = working level	3	m CD	Check validity range h_t/D_{n50}	12.8	ok
Rock Class III: M_{min} (leave open for MA)	1	t	Minimum berm level from construction	6.54	m CD	Check validity range h_t/h	0.66	ok
Rock Class III: M_{max} (leave open for MA)	4	t	Design choice of berm width	24.00	m	*Overload conditions*		
			Design choice of berm level	8.20	m CD	Allowable damage level for overload, N_{od}	4	-
						Highest level of toe for overload with chosen N_{od}	-11.75	m CD
			Required horizontal armour width A_h	34.1	m	Check validity range h_t/D_{n50}	13.1	ok
			Design choice of A_h	34.0	m	Check validity range h_t/h	0.67	ok
						Design choice of toe berm level (0 if no berm)	-12	m CD
						Design choice cotα core below A_h	1.5	-

Section 8.4.3. FR MA standard grading, Class I 6-10 t.

General conditions			**Outcome main parameters**			**Minimum transition level to class II**		
Design wave height H_{sD}	7	m	Wave steepness s_{op}	0.040	-	For H_{sD} at lowest level	-1.8	m CD
Peak period T_p	10.6	s	Relative mass density Δ	1.62	-	For lowest level with according H_s	-2.6	m CD
Overload H_s	8	m	Median mass Class I M_{50}	8.0	t	Design choice of transition for IC (3 rock classes)	-2.6	m CD
Design water level DWL	4	m CD	Nominal diameter Class I D_{n50}	1.44	m	Transition lower class for MA (2 rock classes)	-3.9	m CD
Lowest water level with H_{sD}	1	m CD	Stability number $H_{sD}/\Delta D_{n50}$	3.01	-			
Lowest storm level	0	m CD	**Type of berm breakwater**	**Dyn. stable structure, no breakwater!**		**Crest level (γ_b = 1)**		
H_s at lowest storm level	6.5	m	Number of rock classes for berm	2		If no overtopping criteria, $R_{c\,min}$	12.4	m CD
Mean High Water Spring	2	m CD	Basic recession for H_{sD} (no adaptation)	13.10	m	If no overtopping criteria, $R_{c\,max}$	13.8	m CD
Bottom level of foreshore at toe of structure	-18	m CD	Recession for overload (no adaptation)	20.04	m	For given allowable overtopping, q, γ_{BB}	0.34	
Allowable overtopping q for H_{sD}	10	l/s per m	Nominal diameter Class II, D_{n50}	1.19	m	Required crest level for design conditions	11.37	m CD
Allowable overtopping q for overload	no	l/s per m	Nominal diameter Class III, D_{n50}	No Class III		Required crest level for overload		m CD
Mass density water	1030	kg/m^3				Design choice of crest level	12.50	m CD
Mass density rock	2700	kg/m^3	**Resiliency, berm width and level**					
			Wanted resiliency	50	%	**Check possibility of toe berm at level h_t**		
Choice of rock classes			Resulting Berm width B from resiliency	26.19	m	**Lowest** possible toe level (two layers)	-14.13	m CD
Rock Class I: minimum mass (0-15%)	6	t	Minimum berm width B_{min} from geometry	14.53	m	*Design conditions*		
Rock Class I: maximum mass (85-100%)	10	t	Berm level 0.6 H_{sD}	8.2	m CD	Allowable damage level for H_{sD}, N_{od}	2	-
Rock Class II: minimum mass (0-15%)	3	t	Δw for waves during construction	1	m	*Highest* level of toe for H_{sD} with chosen N_{od}	-9.60	m CD
Rock Class II: maximum mass (85-100%)	6	t	MHWS plus Δw = working level	3	m CD	Check validity range h_t/D_{n50}	8.9	ok
Rock Class III: M_{min} (leave open for MA)		t	Minimum berm level from construction	5.87	m CD	Check validity range h_t/h	0.56	ok
Rock Class III: M_{max} (leave open for MA)		t	Design choice of berm width	27.00	m	*Overload conditions*		
			Design choice of berm level	6.00	m CD	Allowable damage level for overload, N_{od}	4	-
						Highest level of toe for overload with chosen N_{od}	-9.89	m CD
			Required horizontal armour width A_h	42.1	m	Check validity range h_t/D_{n50}	9.2	ok
			Design choice of A_h	42.0	m	Check validity range h_t/h	0.57	ok
						Design choice of toe berm level (0 if no berm)	-10	m CD
						Design choice cotα core below A_h	2	-

Bibliography

Papers written by Van der Meer and/or Sigurdarson can be downloaded from www.vandermeerconsulting.nl or www.icebreak.is.

Baird, W.F. and Hall, K.R. (1983). The design of armour systems for the protection of rubble mound breakwaters, *Proc. Conf. on Breakwaters - Design and Construction*, ICE, pp. 107-119.

Baird, W.F. and Hall, K.R. (1984). The design of breakwaters using quarried stones, *Proc. 19th Conf. Coastal Eng.*, ASCE, pp. 2580-2591.

Baird, W.F. and Woodrow, K. (1987). The design and construction of a mass-armoured breakwater at Hay Point, Australia, *Proc. Berm breakwaters*, ASCE, Seminar on Unconventional Rubble Mound breakwaters, Willis, D.H., Baird, W.F. and Magoon, O.T. (Eds.), pp. 138-146.

Battjes, J. A. and Groenendijk, H. W. (2000). Wave height distributions on shallow foreshores, *Coastal Eng.,* 40, pp. 161-182.

Berm breakwaters (1987). Berm breakwaters: Unconventional Rubble-Mound Breakwaters, ASCE, Willis, D.H., Baird, W.F. and Magoon, O.T. (Eds.).

Bremner, W., Foster, D.N., Miller, C.A. and Wallace, B.C. (1980). The design concept of dual breakwaters and its application to Townsville, Australia, *Proc. 17th Conf. Coastal Eng.*, ASCE, pp. 1898-1908.

Bremner, W., Harper, B.A. and Foster, D.N. (1987). The design and construction of a mass-armoured breakwater at Hay Point, Australia, *Proc. Berm breakwaters*, ASCE, Seminar on Unconventional Rubble Mound breakwaters, Willis, D.H., Baird, W.F. and Magoon, O.T. (Eds.), pp. 147-218.

Bruce, T., Van der Meer, J.W., Franco, L. and Pearson, J.M. (2009). Overtopping performance of different armour units for rubble mound breakwaters, *Special Issue of Coastal Eng.*, 56, pp. 166-179.

Bruun, P. (1985). Design and construction of mounds for breakwaters and coastal protection, Elsevier.

Bruun, P. and Johannesson, P. (1976). Parameters affecting the stability of rubble mounds, *J. Waterway Port Coastal Ocean Eng.*, ASCE, 102 (2), pp. 141-164.

Burcharth, H.F. and Frigaard, P. (1987). On the stability of berm breakwater roundheads and trunk erosion in oblique waves, *Proc. Berm breakwaters*, ASCE, Seminar on Unconventional Rubble Mound breakwaters, Willis, D.H., Baird, W.F. and Magoon, O.T. (Eds.), pp. 55-72.

CLASH (2004). Crest Level Assessment of coastal Structures by full scale monitoring, neural network prediction and Hazard analysis on permissible wave overtopping, Fifth Framework Programme of the EU, Contract n. EVK3-CT-2001-00058, www.clash-eu.org.

Coastal Engineering Manual (2006). Coastal Engineering Research Centre, Report No. EM 1110-2-1100, US Army Corps of Engineers, Washington DC, USA.

Colleter, G., Van Staden, A., Blacka, M. and Louys, J. (2011). Mackay Breakwater Adaptability, *Proc. Coasts and Ports 2011*, Australia.

Espedal, T.G. and Lothe, A.E. (1994). The Mortavika breakwater: a harbour exposed to severe weather, *Proc. Symposium on Strait Crossings*, Balkema, pp. 877-883.

EN 13383 (2002). EN13383-1:2002 Armourstone. Specification, EN 13383-2:2002 Armourstone. Test methods.

EurOtop (2007). Wave overtopping of sea defences and related structures – Assessment Manual, Allsop, N.W.H., Pullen, T. Bruce, T., Van der Meer, J.W., Schüttrumpf, H. and Kortenhaus. A., www.overtopping-manual.com.

EurOtop (2016). Manual on wave overtopping of sea defences and related Structures. An overtopping manual largely based on European research, but for worldwide application, Allsop, N.W.H., Bruce, T., De Rouck, J., Kortenhaus, A., Pullen, T., Schüttrumpf, H., Troch, P., Van der Meer, J.W. and Zanuttingh, B., www.overtopping-manual.com.

Foster, D.N., McGrath, B.L. and Bremner, W. (1978). Rosslyn Bay breakwater, Queensland, Australia. The design of breakwaters using quarried stones, *Proc. 16th Conf. Coastal Eng.*, ASCE, pp. 2086-2103.

Gilman, J.F. (1987). Performance of a berm roundhead in the St. George breakwater system, *Proc. Berm breakwaters*, ASCE, Seminar on Unconventional Rubble Mound breakwaters, Willis, D.H., Baird, W.F. and Magoon, O.T. (Eds.), pp. 219-228.

Gilman, J.F. (1999). St. George Harbour – Berm breakwater construction considerations, *Proc. Breakwaters '99*, ASCE, First International Symposium on Monitoring Breakwaters, pp. 107-121.

Gourlay, M.R. (1996). History of coastal engineering in Australia, In History and Heritage of Coastal Engineering, ASCE, ed. Kraus, N.C., pp. 1-188

Hall, K.R., Baird, W.F. and Rauw, C.I. (1983). Development of a wave protection scheme for a proposed offshore runway extension, *Proc. Coastal Structures '83*, ASCE, pp. 157-170.

Hall, K.R. and Kao, J.S. (1991). A study of the stability of dynamically stable breakwaters, *Canadian J. of Civil Engineering*, Vol. 18, pp. 916-925.

Johnson, R.A., McIntyre, P.T. and Hooper, G.R. (1999). Mackay Small Craft Harbour Breakwater, A Case Study in Reshaping Berm Breakwaters, *Proc. Coasts and Ports 1999*, Australia.

Juhl, J. and Jensen, O.J. (1995). Features of berm breakwaters and practical experience, *Proc. COPEDEC*, pp. 1307-1320.

Juhl, J., and Sloth, P. (1998). Berm breakwaters - influence of stone gradation, permeability and armouring, *Proc. 26th Conf. Coastal Eng.*, ASCE, pp. 1394-1406.

Le H.T., Van der Meer, J.W. and Verhagen, H.J. (2014). Wave overtopping simulator tests on Vietnamese seadikes, *Coastal Eng. Journal, 56*, No. 3, 1450017.

Lissev, N. (1993). Influence of the core configuration on the stability of berm breakwaters. Experimental model investigations, Report R-6-93, University of Trondheim, the Norwegian Institute of Technology.

Lothe, A.E. and Espedal, T.G. (1994). Design and Construction of Berm Breakwaters - Norwegian Experience, *Proc. Int Coastal Symposium, Iceland*, pp. 737-748.

Lothe, A.E. and Lindefjeld, L. (2005). The Mortavika breakwater – 12 years later, *Proc. Second Int. Coastal Symposium in Iceland.*

Lykke Andersen, T. (2006). Hydraulic response of rubble mound Breakwaters. Scale effects – berm breakwaters, Ph.D. Thesis, Department of Civil Engineering, Aalborg University, Denmark.

Lykke Andersen, T., Skals, K.T. and Burcharth, H.F. (2008). Comparison of homogenous and multi-layered berm breakwaters with respect to overtopping and front slope stability, *Proc. 31th Conf. Coastal Eng.*, ASCE.

Lykke Andersen, T. and Burcharth, H.F. (2010). A new formula for front slope recession of berm breakwaters, *Coastal Eng.*, 57, pp. 359–374.

Lykke Andersen, T., Van der Meer, J.W., Burcharth, H.F. and Sigurdarson, S. (2012). Stability of hardly reshaping berm breakwaters, *Proc. 26th Conf. Coastal Eng.*, ASCE.

MAST I (1993). G6 Coastal Structures, Final report under EU-MAST I research contract.

MAST II (1997). Berm Breakwater Structures, Final Report under MAST-CONTRACT MAS2-CT94-0087.

Moghim, N., Tørum, A., Arntsen, Ø. and Brørs, B. (2009). Stability of berm breakwaters for deep water, *Proc. 4th SCACR.*

Moghim, M.N., Shafieefar, Tørum, A. and Chegini, V. (2011). A new formula for the sea state and structural parameters influencing the stability of homogeneous reshaping berm breakwaters, *Coastal Eng.*, 58, pp. 706-721.

Myhra, H.A. (2005). Berm breakwaters. Influence of construction method and storm duration on stability, MSc-thesis, Norwegian University of Science and Technology, NTNU.

Nurmohamed, S., Steendam, G.J. and Van der Meer, J.W. (2006). Weight and stability assessment of single layers of orderly placed or pitched natural rock, *Proc. 30th Conf. Coastal Eng.*, ASCE, pp. 4815-4827.

PIANC (2003). State-of-the-Art of designing and constructing berm breakwaters, MarCom Report of WG 40.

Priest, M.S., Pugh, J.W. and Sigh., R. (1964). Seaward profile for rubble mound breakwaters, *Proc. 9th Conf. Coastal Eng.*, ASCE, pp. 553-559.

Rock Manual (1991). *Manual on the use of rock in coastal and shoreline engineering*, CIRIA Special Publication 83, CUR Report 154.

Rock Manual (2007). *The Rock Manual. The use of rock in hydraulic engineering*, CIRIA, CUR, CETMEF.

Salauddin, Md, Broere, A., Van der Meer, J.W., Verhagen, H.J. and Bijl, E. (2015). A new symmetrical unit for breakwater armour: first tests, *Proc. Coastal Structures 2015*, ASCE.

Sigurdarson, S. and Viggosson, G. (1994). Berm breakwaters in Iceland, practical experiences, *Proc. of Hydro-Port' 94*, PHRI, pp. 651-671.

Sigurdarson, S., Viggosson, G., Benediktsson, S. and Smarason, O.B. (1996). Berm breakwaters, tailor-made size graded structure, *Proc. 11th Int. Harbour Congress*, pp. 285-296.

Sigurdarson, S., Viggosson, G., Benediktsson, S., Einarsson, S. and Smarason, O.B. (1998). Berm breakwaters, fifteen years experience, *Proc. 26h Conf. Coastal Eng.*, ASCE, pp. 1407-1420.

Sigurdarson, S., Bjornsson, B.J., Skulason, J., Viggosson, G. and Helgason, K. (1999). A berm breakwater on a weak soil, extension of the port of Hafnarfjordur, *Proc. COPEDEC*, pp. 1395-1406.

Sigurdarson, S., Smarason, O.B. and Viggosson, G. (2000). Design Considerations of Berm Breakwaters, *Proc. 27th Conf. Coastal Eng.*, ASCE, pp. 1610-1621.

Sigurdarson, S., Jacobsen, A. Smarason, O.B., Bjordal, S., Viggosson, G., Urrang, C. and Tørum, A. (2003). Sirevag berm breakwater, design, construction and experience after design storm, *Proc. Coastal Structures 2003*, ASCE, pp. 1212-1224.

Sigurdarson, S., Loftsson, A., Lothe, A.E., Bjertness, E. and Smarason, O.B. (2005-a). Berm breakwater protection for the Hammerfest LNG Plant in Norway - Design and Construction, *Proc. Coastlines, Structures and Breakwaters*, ICE, Thomas Telford, pp. 349-362.

Sigurdarson, S., Viggosson, G. and Smarason, O.B. (2005-b). Recent berm breakwater projects, *Proc. Second Int. Coastal Symposium in Iceland.*

Sigurdarson, S., Smarason, O.B., Viggosson, G. and Bjordal, S. (2006). Wave height limits for the statically stable Icelandic-type berm breakwater, *Proc. 30th Conf. Coastal Eng.*, ASCE, pp. 5046-5058.

Sigurdarson, S., Gretarson S.A. and Van der Meer J.W. (2008). The Icelandic-type berm breakwater for large design wave heights, *Proc. COPEDEC VII.*

Sigurdarson, S. and Van der Meer, J.W. (2011). Front slope stability of the Icelandic-type berm breakwater, *Proc. Coastal Structures 2011*, ASCE, pp. 435-446.

Sigurdarson, S. and Van der Meer, J.W. (2012). Wave overtopping at berm breakwaters in line with EurOtop, *Proc. 33th Conf. Coastal Eng.*, ASCE.

Sigurdarson, S., Mocke, R., Primmer, P. and Gretarsson, S. (2012). Armourstone for the Icelandic-type Berm Breakwater. *Proc. PIANC COPEDEC 2012.*

Sigurdarson, S. and Van der Meer, J.W. (2013). Design of berm breakwaters, recession, overtopping and reflection, *Proc. Coasts, Marine Structures and Breakwaters 2013*, ICE.

Sigurdarson, S., Van der Meer, J.W., Bijl, E., Yang Sihan, Tang Qiaoliang, Zhang Xiaoqiang, James KS Goh and Heijboer, D. (2014). Icelandic-type berm breakwater for the Hambantota artificial island revetment, application of geometrical design rules, *Proc. 34th Conf. Coastal Eng.*, ASCE.

Sigurdarson, S. and Van der Meer, J.W. (2015). Design and construction of berm breakwaters, *Proc. Coastal Structures 2015*, ASCE.

Smarason, O.B., Sigurdarson, S. and Viggosson, G. (2000). Quarry yield prediction as a tool in breakwater design, Keynote lectures NGM-2000, Finish Geotechnical Society.

Stewart, T.P., Newsberry, S.D., Latham, J.-P. and Simm, J.D. (2003). Packing and voids for rock armour in breakwaters, HR Wallingford report SR 621.

Sveinbjörnsson, P.I. (2008). Stability of Icelandic type Berm Breakwaters, MSc-thesis Delft University of Technology.

Thompson, D.M. and Shuttler, R.M. (1975). Riprap design for wind wave attack. A laboratory study in random waves, HRS Wallingford, Report EX 707.

Thomsen, J.B., Røge, M.S., Christensen, N.F., Lykke Andersen, T. and Van der Meer, J.W. (2014). Stability of hardly reshaping berm breakwaters exposed to long waves, *Proc. 34th Conf. Coastal Eng.*, ASCE.

Tørum, A. (1998). On the stability of berm breakwaters in shallow and deep water, *Proc. 26th Conf. Coastal Eng.*, ASCE, pp. 1435-1448.

Tørum, A., Krogh, S.R., Bjørdal, S., Fjeld, S., Archetti, R. and Jacobsen, A. (1999). Design criteria and design procedures for berm breakwaters, *Proc. Coastal Strucstures '99*, pp. 331-341.

Tørum, A., Kohen, F. and Menze, A. (2003). On berm breakwaters. Stability, scour, overtopping, *Coastal Eng.*, 49, pp. 209-238.

Tørum, A., Løset, S. and Myhra, H. (2005). Sirevåg berm breakwater – comparison between model and prototype reshaping, *Proc. Second Int. Coastal Symposium in Iceland.*

Van der Meer, J.W. (1985). Stability of rubble mound revetments and breakwaters under random wave attack, *Proc. Breakwaters 1985*, ICE, pp. 141-154, Discussion pp. 191-202

Van der Meer, J.W. (1987-a). Stability of breakwater armour layers - Design formulae, *Coastal Eng.*, 11, pp. 219-239.

Van der Meer, J.W. (1987-b). Application of computational model on berm breakwater design, *Proc. Berm breakwaters*, ASCE, Seminar on Unconventional Rubble Mound breakwaters, Willis, D.H., Baird, W.F. and Magoon, O.T. (Eds.), pp. 92-103.

Van der Meer, J.W. (1988-a). Rock slopes and gravel beaches under wave attack, Doctoral thesis, Delft University of Technology.

Van der Meer, J.W. (1988-b). Deterministic and probabilistic design of breakwater armour layers, *J. Waterway Port Coastal Ocean Eng.*, ASCE, 114, No. 1, pp. 66-80.

Van der Meer, J.W. (1990). Static and dynamic stability of loose materials, Coastal Protection, Pilarczyk (Ed.), Balkema, pp. 157-195.

Van der Meer, J.W. (1992). Stability of the seaward slope of berm breakwaters, *Coastal Eng.*, 16, pp. 205-234.

Van der Meer, J.W. (1994). Conceptual design of berm breakwaters, *Proc. Int. Coastal Symposium, Iceland*, pp. 801-811.

Van der Meer, J.W. (1995). Conceptual design of rubble mound breakwaters, Advances in Coastal and Ocean Engineering, Volume 1, Liu, P.L.F. (Ed.), pp. 221-315.

Van der Meer, J.W. (2014). Simulators as hydraulic test facilities at dikes and other coastal structures, *Series of Coastal and Ocean Engineering Practice*, Vol. 2, Design of Coastal Structures and Sea Defences, Kim, Y.C., (Ed.), pp. 1-22.

Van der Meer, J.W. and Pilarczyk, K.W. (1984). Stability of rubble mound slopes under random wave attack, *Proc. 19th Conf. Coastal Eng.*, ASCE. pp. 2620-2634.

Van der Meer, J.W. and Pilarczyk, K.W. (1986). Dynamic stability of rock slopes and gravel beaches, *Proc. 20th Conf. Coastal Eng.*, ASCE, pp. 1713-1728.

Van der Meer, J.W. and Koster, M.J. (1988). Application of computational model on dynamic stability, *Proc. Breakwaters '88*, ICE, pp. 333-342.

Van der Meer, J.W. and Veldman, J.J. (1992). Singular points at berm breakwaters: scale effects, rear, roundhead and longshore transport, *Coastal Eng.*, 17, pp. 153-171.

Van der Meer, J.W., Nurmohamed, S., Philipse, L.A., Steendam, G.J. and Wouters, J. (2005). Stability Assessment of single layers of orderly placed and of pitched natural rock, *Proc. Second Int. Coastal Symposium*, Iceland.

Van der Meer, J.W., Bernardini, P., Snijders, W. and Regeling, H.J. (2006). The wave overtopping simulator, *Proc. 30th Conf. Coastal Eng.*, ASCE, pp. 4654-4666.

Van der Meer, J.W., Hardeman, B., Steendam, G.J. Schüttrumpf, H. and Verheij, H. (2010). Flow depths and velocities at crest and inner slope of a dike, in theory and with the Wave Overtopping Simulator, *Proc. 32th Conf. Coastal Eng.*, ASCE.

Van der Meer, J.W., Thornton, C. and Hughes, S. (2011). Design and operation of the US Wave Overtopping Simulator, *Proc. Coastal Structures 2011*, pp. 155-166.

Van der Meer, J.W. and Sigurdarson, S. (2014). Geometrical design of berm breakwaters, *Proc. 34th Conf. Coastal Eng.*, ASCE.

Van Gent, M.R.A, Smale, A.J. and Kuiper, C. (2003). Stability of rock slopes with shallow foreshores, *Proc. Coastal Structures 2003*, ASCE, pp. 100-112.

Van Gent, M.R.A. and Van der Werf, I.M. (2014). Rock toe stability of rubble mound breakwaters, *Coastal Eng.*, 83, pp. 166-176.

Viggosson, G., (1990). Rubble Mound Breakwaters in Iceland, *Journal of Coastal Research*, Special Issue 7, pp. 41-61.

Zanuttigh, B and Van der Meer, J.W. (2008). Wave reflection from coastal structures in design conditions, *Coastal Eng.*, 55, pp. 771-779.

About the Authors

Dr **Jentsje van der Meer** is a well-known expert in appraisal, design and testing of breakwaters and coastal structures, including levees, dikes, embankments, seawalls, breakwaters, groynes, revetments, shingle beaches and river dikes. His work on rubble mound structures has been included in all manuals all over the world. He has worked at Delft Hydraulics, now Deltares, a well-known institute on specialised consulting and research of water related issues, for 16 years. For ten years he had a position at Infram International, a private consultant for infrastructure appraisal and management, where he exploited his experience in specialized consultancy and research. In 2007 he started his own company, Van der Meer Consulting B.V., on Coastal Engineering Consultancy & Research. In 2014, he became a part time professor of Coastal Structures and Ports (0.2 fte) at UNESCO-IHE, Delft, The Netherlands, with also a 0-fte position at Delft University of Technology.

In addition to the applied research in the first part of his career Dr Van der Meer has been project manager or advisor in many projects on design of all kinds of coastal structures all around the world - many of which have since been constructed. He has also performed applied research for the Dutch government (various departments) and the European Union. Dr Van der Meer has since widened his work related to wave-structure interaction; such as wave run-up and wave overtopping at dikes, including the strength of these structures under wave attack. He developed a Dutch guideline on wave run-up and overtopping at dikes.

The EU-project, CLASH, was initiated by him and included research institutes like HR Wallingford and Delft Hydraulics/Deltares. He was co-author of the *EurOtop Manuals on Wave Overtopping* (2007 and 2016), which brings together the knowledge of the UK, the Netherlands, Germany and Belgium.

Dr Van der Meer is and has been chairman or member of a large number of national and international committees with respect to safety assessment and the design of coastal structures. He has published about two hundred papers in international journals, proceedings and books. He won the Halcrow Premium in 1987 and the T.K. Shieh award in 1992, both granted by the UK Institution of Civil Engineers. Dr Van der Meer has also contributed to many short courses and post-doc courses.

His latest developments are primarily the Hydraulic Simulators; like the Wave Overtopping Simulator, the Wave Run-up Simulator, and the Wave Impact Generator (Deltares). These are unique devices that test the strength of real dikes *in-situ* for impacting, up-rushing and overtopping waves. These simulators have been designed and constructed, and have been used to test the strength of various existing dikes or levees in the Netherlands from 2007-2014. Other Simulators have been designed for Vietnam and for the US.

More information, including many downloads of papers and data on: www.vandermeerconsulting.nl.

Sigurdur Sigurdarson has over 30 years of experience as a coastal and harbour engineer, both in Iceland as well as internationally, working on breakwater projects in four continents. His main emphasis has been on coastal structures, including breakwaters, revetments and groynes. He has been involved in all aspects from planning, establishment of environmental load and design criteria, design, model testing and quarry evaluation, through to tendering, construction management, supervision of construction and quarrying, as well as performance monitoring. Through a number of breakwater projects, he has developed and introduced the Icelandic-type berm breakwater.

Sigurdarson started working for the Icelandic Harbour Authority in 1984, where he joined a small team managing nearly all harbour and coastal projects in Iceland. This team has been responsible for nearly all aspects of the projects, wave measurements, wave modelling, bathymetric survey, physical model testing, designing, tendering, supervising and monitoring. Although there have been institutional changes over the years (the Icelandic Harbour Authority merged into Icelandic Maritime Administration and recently into Icelandic Road and Coastal Administration), the core of the team has remained roughly the same.

The development and design of rubble mound breakwaters and revetments has been an important part of Sigurdarson's work. This has involved armourstone quarry assessment, material specifications, supervision of quarrying, advising contractors in blasting for armourstone, as well as design. Sigurdarson has designed rubble mound breakwaters in over fifty harbours in Iceland and an even higher number of shore protection projects.

The development of the Icelandic-type berm breakwaters opened opportunities to work on projects abroad; the first two projects being exposed breakwaters in Norway - the Sirevåg breakwater with a design wave height of 7.0 m and the Hammerfest breakwater or berm revetment, protecting an LNG plant against overtopping, with a design wave height of 7.5 m. Gradually more international projects followed.

Sigurdarson established the IceBreak Consulting Engineers, which specialises in breakwaters and armourstone quarrying, in 2010. An important factor in Sigurdarson's professional development has been the cooperation with Jentsje van der Meer, the first author of this book. It started in international working groups and research projects on berm breakwaters, developed though the writing of papers on berm breakwaters as well as in consulting projects and eventually resulted in the writing of the present book.

More information on: www.icebreak.is.